LIFE
IN THE
UNIVERSE

JEAN

HEIDMANN

LIFE

IN THE

UNIVERSE

McGraw-Hill, Inc.

New York St. Louis San Francisco Auckland Bogotá
Caracas Hamburg Lisbon London Madrid
Mexico Milan Montreal New Delhi Paris
San Juan São Paulo Singapore
Sydney Tokyo Toronto

English Language Edition

Translated by Isabel A. Leonard
in collaboration with
The Language Service, Inc.
Poughkeepsie, New York

Typography by AB Typesetting
Poughkeepsie, New York

Library of Congress Cataloging-in-Publication Data

Heidmann, Jean.
   [*La Vie dans l'univers*. English]
   Life in the Universe/Jean Heidmann.
      p. cm. — (The McGraw-Hill *HORIZONS OF SCIENCE* series)
   Translation of: *La Vie dans l'univers*.
   Includes bibliographical references.
   ISBN 0-07-027887-3
   1. Life on other planets. 2. Interstellar communication.
   I. Title. II. Series.
QB54.H39713 1992
574.999—dc20   91-32437

The original French language edition of this book
was published as *La Vie dans l'univers*, copyright © 1989,
Questions de science series
Hachette, Paris, France.
Series editor, Dominique Lecourt

TABLE OF CONTENTS

INTRODUCTION

Are there extraterrestrial beings? Are they intelligent? More intelligent than we, or better organized, or even civilized? Do they send us messages? Will we be able to communicate with them? Very soon, perhaps as soon as 1992, we will be able to answer these fascinating, ever-tantalizing and irksome questions.

The reader of this book will discover an entirely new science that is officially only ten years old: bio-astronomy, the study of life in the Universe in all its forms. Jean Heidmann presents the wealth of his results, which are already impressive; he shows how numerous research avenues are coming together, from cosmology to most fundamental theoretical physics (relativity and quantum mechanics), as well as geology, paleontology, and molecular biology. It is a perfect example of one of the most up-to-date methods of scientific progress today: through cross-fertilization, research borrows concepts, methods and experimental techniques.

In radio astronomy, technology is of particular importance: the reader will go to the very heart of one of the most ambitious projects ever mounted by NASA: the construction of a superreceiver that will increase our capacity to listen to the Universe by a factor of ten thousand all in one leap! But the reader will also realize the extent to which technology, when developed to serve scientific investigation, is rigorously and meticulously dependent on strategies of knowledge and on the theoretical framework that guides and constricts it. After reading Jean Heidmann, there can be no doubt that today more than ever science is an exciting, intellectual adventure. The prophecies of doom of those who see "technoscience" as a threat to the planet and the human species pale in the face of his pioneering enthusiasm.

Of course, our attention will certainly be drawn to the ultimate prospects of this research, which are not only to record the already attested existence of organic molecules and other forms of primitive life in the Universe in order to inventory them, but to detect signals that will finally give us the answer to the great question: Are there intelligent beings elsewhere than on Earth?

We will see the precise scientific and technical terms in which the problem is formulated today. But what gives it its deep resonance, what fires the imagination of some and elicits the irony of others, is the underlying philosophical question. A very old question, indeed, and a most thorny one at that. It will be worth while, perhaps on the eve of its disappearing from the conscious (if not subconscious) human mind, to retrace its history back in time, a history replete with both tragedy and feats of daring.

In 1686, one year before the publication in Latin of Isaac Newton's *Philosophiae Naturalis Principia Mathematica* [Mathematical principles of natural philosophy], Bernard de Fontenelle, a young writer from

Rouen noticed by Pierre Bayle, addressed to the educated public of his day the *Entretiens sur la pluralité des mondes* [Conversations on the plurality of worlds], a work of iconoclastic inspiration confirmed shortly afterward by his *Histoire des oracles* [History of oracles], which earned him the violent hostility of Père Lachaise, the king's confessor. Eleven years later, he was elected to the Academy of Sciences, whose highly respected permanent secretary he became after 1699.

In the course of six evenings, a marquise, astounded but enthusiastic, received an astronomy lesson from the philosopher. The lesson owed a great deal to Copernicus, but especially to Descartes. As early as the first evening there was mention of "a certain Athenian madman who took it into his head that every vessel reaching the port of Piraeus belonged to him"; this was followed by the aphorism, which summarizes the spirit of the *Entretiens*, "Our own folly is to believe that all of Nature, without exception, is intended for our use." This served as a pretext to invoke the celebrated theory of vortices that Descartes had expounded, particularly in the third part of *Principia Philosophiae* [Principles of philosophy] (1644), to rebut the notion of a center of the Universe and to advance the idea that "the fixed stars are so many Suns, each of which illuminates a world," after having affirmed on the second evening, as if self-evident, that "the Moon is an inhabited Earth."

To the marquise who exclaimed, overcome by dizziness, "But this Universe is so big, I get lost in it; I don't know where I am—now I know nothing at all," the philosopher replied: "Now that the celestial vault has been given an infinitely greater expanse and depth (...), I feel as if I can breathe more freely, that my realm is greater, and that assuredly magnificence of the Universe is something else entirely." To his incredulous lady companion, unable to believe in the existence of intelligent beings on other planets, he tells the story of the Parisian bourgeois who, having never left his city, gazed at the suburb of Saint-Denis from the topmost tower of Notre Dame and was ready to swear that Saint-Denis was uninhabited since he could not see the inhabitants. Clearly, Fontenelle had in mind the posthumous work by that most turbulent of the 17th century libertines, Cyrano de Bergerac (1619–1655), published in 1657 under the title *Histoire comique contenant les états et empires de la lune* [Comical history containing the states and empires of the Moon]. The text had been expurgated for publication and the title distorted by a cautious but unscrupulous friend. Amusing and even farcical as the plot may be, *L'autre monde ou les états et empires de la lune* [The other world or the states and empires of the Moon] was not in fact a "comic history" but rather a scathing satirical history and a spirited plea for materialism: the "other world" is no longer the "beyond," the prospect of Salvation, but another existing world, "parallel" with our own!

It is to this too often forgotten tradition that Fontenelle unquestionably belongs. His playful humor ought not to conceal the gravity of the question he raised. The Church was undeceived: it saw, glittering dangerously, the resurgence of a philosophical thesis a thousand times accursed, with which there was no possible compromise.

Giordano Bruno (1548–1600) had been arrested by the Inquisition in Venice in 1592, then burned at the stake in Rome on February 17, 1600, after eight appalling years of detention and solitude, not only for having taken the heliocentric theory of Copernicus literally without oratorical precautions and then having advanced new arguments, ingenious but more popular than scientific, in favor of the movement of the Earth, but perhaps above all for having affirmed the infinity of the Universe, still not accepted by the Polish astronomer, as well as the infinite plurality of worlds. Bruno sincerely believed that this would redound to the greater glory of God. But his judges, not without reason, saw it as an intolerable twisting of the essential dogmas of the Christian faith. Was not the act of creation unique? And did not God create us, and us alone, in his image in this unique world?

The inquisitors readily discerned the thread connecting the concept of the old Dominican of Naples with an immemorial philosophical tradition— attacked, mocked, and calumniated since the beginning of time by all the forces that fashioned the intellectual order of the West.

11

As far as we know, the Ionian philosopher, Anaximander of Miletus, was the first to speak of a plurality of worlds, in the 6th century B.C. If, as he believed in his cosmogony, the Infinite—or, more precisely, the Unlimited (in Greek: *apeiron*)—was this indeterminate "element" from which sprang all the determinate things that formed our world contained within the boundaries of the heavens, several worlds could exist in it simultaneously which, as he so splendidly put it, "are born and perish within an eternal and ageless infinity."

Anaximenes (550–480 B.C.) somewhat later took up the gist of this thesis by transfering the properties of infinity to air.

One might think that these charming Ionian images, in their touching archaism, were buried in far too remote a past to disturb the late 17th-century mind. Perhaps so. But Epicurus (341–270 B.C.) dusted off these old theories, restored their full sparkle and endowed them with an ethical and political virulence that never tarnished throughout history. Epicurus is known for having renewed and reformed atomic physics founded by Leucipus of Miletus (ca. 460–370 B.C.) and Democritus of Abdera (5th century B.C.), but he also restored to honor the idea of a multiplicity of worlds drawing their matter from infinity. To the cosmologies of Plato and Aristotle, who believed the world to be identical to the entire Universe, and to that of the stoics, so powerful in the 3rd century B.C., who saw it as a huge individual organism, Epicurus boldly opposed the idea that the whole is made up of an infinite number of atoms in the infinite magnitude of vacuum. As he saw it, therefore, a world is only a portion that becomes detached from the infinite and preserves an always precarious order for a brief span of time; hence, there must be an infinity of worlds. What is more, in this infinity, atoms can travel from one world to another. But, according to Epicurus, there was no reason to think that these worlds were of a unique type and that, for example, the same living beings would inhabit them. Quite the contrary, given the infinite diversity of the seeds from which they are generated, there is every reason to imagine that these beings would be quite different.

Adventurous speculations? Dogmatic rationalists of every age have always reproached him in the name of science, even the highly uncertain science of his time. But here are the conclusions Epicurus drew: the world is conceived as a whole only to be the place where human beings make their home. Thus, the dominant physical doctrines are secretly inspired by an anthropocentrism which clearly appears to be a defect in thinking. This anthropocentrism thus feeds the correlative notion of a "demiurge," the creator of the world and its inhabitants, who live in terror of the punishment they are supposed to receive at his hands after death.

By affirming the infinity of the Universe and the plurality of worlds, Epicurus sought to liberate people from fear of the gods and the related superstitions (divination, prophecies, belief in destiny and the like). And so the philosophical fate of epicureanism, which found in Lucretius its somber and majestic hero in Cicero's Rome of the 1st century B.C., was sealed. All those daring to take their inspiration from it in the future would be exposed to public condemnation as irreligious and thus necessarily immoral and debauched.

Although atomism was reborn in 17th-century physics and, after profound changes, came in the course of time to be the successful science we know today, the idea of a plurality of worlds was banished from science and philosophy. Fontenelle notwithstanding, Descartes did not give it a thought, Leibniz evoked it only to relegate it among the kinds of "possible worlds" of which God chose the best. At the same time, the philosophy of Spinoza, radically hostile to anthropocentrism, subscribed to the Epicurean tradition by linking it, in turn, with the idea of a Creator it rejects. But Spinoza, like Epicurus, is one of the accursed philosophers.

Kant, in his "Critique of Pure Reason" (1781), fleetingly evokes the figure of a rational being living on another planet whose sensitivity is not, like ours, structured in accordance with *a priori* forms of space and time, in order to point up our finiteness, and forbids us from speculating about such kinds of illusory questions that lead reason astray when it claims to access realms of knowledge by freeing itself of any possible experience.

One can therefore say that the dominant forms of Western rationalism have rejected the thesis of a plurality of worlds as well as the question of the existence of intelligent extraterrestrial life for philosophical reasons that barely conceal their religious, ethical, and political motives.

This is why, it seems to me, when science reawakened its reality in the 19th century, it had to take refuge in an unclassifiable form of thinking and writing: science fiction. Guy Lardreau has very clearly shown how the highly uneven works published under this label since the last century must not be evaluated as "great" literature so called, much as one might wish to confer dignity and a patent of nobility on the genre. The point about science fiction is not whether it belongs to a particular literary genre, even though sometimes felicity of style smiles upon it in a way that many a novelist might envy. Nor is it a form of popular science, however numerous and precise in some instances the references to science and scientific data. Neither does it have anything to do with the works of pseudoscience: writings on parapsychology, numerology, sophrology, and others of the same ilk that are today's best sellers.

The lively and successful works of science fiction (think of H. G. Wells cited by our author, or Isaac Asimov, or Philip K. Dick) actually gather the philosophical questions that are opened up by the perspectives of scientific research, when it has to take the risk of rational or at least reasonable fiction. To these questions, excluded by the dominant mode of philosophizing, science fiction gives a voice and, through images and settings that are admittedly too often anthropomorphic, carries them to their logical extreme. It is not hard to see why science fiction works enjoy a huge and continuing popular success, which could have been that of philosophy, in the abstract register, if it had not for decades set itself apart from the living activity of research. It is hardly surprising that science fiction has elicited the contempt, often fiercely expressed, of literary and intellectual authorities. It challenges fundamental beliefs, even if from time to time it moralizes in the American style.

So here is one of the world's most powerful scientific institutions spending considerable sums on a project that just yesterday would have been mocked as science fiction. And the Soviets are not far behind. Once again we see how wrong we would have been.

But the great moment approaches. Will our narcissism withstand this new blow, if inflicted? When you read the dazzling text of Jean Heidmann, you start dreaming of a great wind of freedom!

Dominique LECOURT

I

THE IDEA OF

EXTRATERRESTRIAL

LIFE

The idea that intelligent life might exist elsewhere than on Earth is not new. As far back as we go in history, it has nourished many of the most extravagant and captivating dreams of humankind. But it found its first scientific foundation in the 17th century after the theoretical work of Nicholas Copernicus (1473–1543) and, in particular, the observations of Galileo (1564–1642). Indeed, from the moment Mercury, Venus, Mars, Jupiter, and Saturn were clearly seen to be globes, from the moment mountains and cirques were discovered on the Moon, it was valid to wonder whether objects that seemed so similar to Earth could be peopled with living, conscious beings.

This question, and the speculation it brought in its wake, came once more to life in the last century when the great telescopes like that of the Meudon Observatory (1877) were built and technology made it possible to enlarge the diameter of lenses (to over 3 feet in the United States), and lengthen their focus; images of a clarity and precision that could not have been dreamed of twenty years earlier were obtained.

It was Mars, the red planet, that quickly became the privileged focus of research and questions about life in the Universe. This planet revolved on its own axis in nearly 24 hours, give or take a few minutes, just like the Earth, and this rotation occurred on an axis likewise inclined in relation to its orbit just like the Earth, give or take a degree. Moreover, shapes were discovered which looked like desert continents, pink, gray, and brown; there were expanses that looked like seas; clouds and sandstorms, and above all, two polar ice caps in the north and south that melted in the summer and froze in the winter.

From the work of pioneers such as Jules Janssen (1824–1907), founder of the Meudon Observatory, the conclusion was reached somewhat too hastily that there was water vapor on Mars. Water vapor suggests the possibility of life, but the existence of this water vapor would not by itself have aroused the craze for Mars that lasted for decades. Other, truly staggering observations, suggested that not only was there life on this planet, but even intelligent and organized life!

We must remember that every two years Mars moves close to the Earth and opposite the Sun. Illuminated head-on, it can easily be seen with the naked eye in the middle of the night; it is more luminous than ordinary stars. A strange and huge disquieting spectacle: there it is, shining with a powerful, steady red glow, without twinkling. Since its orbit is fairly elongated, its approaches to Earth come more or less close. In 1877 Mars came very close to the Earth, even closer than in 1988. So astronomers aimed their brand new telescopes just born of the latest technology. Giovanni Schiaparelli (1835–1910), director of the Milan Observatory and a meticulous astronomer of the highest competence, now discovered a complicated network of dark lines which he interpreted as canals joining the dark patches that were then believed to be seas.

Two years later, Schiaparelli confirmed his observation but was far more cautious in his interpretation. Yet another surprise came in 1891–1892: he saw the "canals" split in two during certain (Martian) seasons. Where in previous days he had seen only one line, he discovered that the canal was now formed by two straight parallel lines a few hundreds of miles apart! He could hardly believe his eyes, but the evidence was there; he found over twenty examples of such doubling. The conclusion appeared ineluctable: "Everything suggests that there is a special organization of the planet Mars." One of his colleagues, carried away by enthusiasm, wrote in the *Times* that "the inhabitants of Mars must be engaged in engineering works of vast scope."

And so the great saga of the Martians was launched in 1892; it proved to be a lasting one. The controversy over the canals of Mars was to last for thirty years, until in 1909 the Frenchman Eugène Antoniadi (1870–1944), looking through the all-powerful telescope at Meudon, observed that Schiaparelli's canals did have some basis in observation but were not real canals or even straight or double lines. More powerful and precise observation revealed them to be infinitely irregular natural structures.

Space exploration has now revealed the "truth" about the Martian canals: probes have shown that on the surface of Mars there is a great canyon 4000 kilometers [2500 miles] long by 250 kilometers [150 miles] wide and 9000 kilometers [5500 miles] deep! The Grand Canyon would be a tiny crack compared to this gorge. Was this a gigantic rip in the crust of the planet, caused by the upheavals of giant volcanos? As for the famous Martian "seas," there is nothing aquatic about them; this is only how they appear as a result of the dark basalt rock swept by violent winds that raise the reddish dust covering the deserts of the planet. It is true that space probes have also revealed the existence of dry beds of ancient rivers over 15 kilometers [10 miles] wide. We will come back to the importance of this discovery later.

An entire body of science fiction literature has developed around Schiaparelli's conjectures. The most interesting book is doubtless *The War of the Worlds* by H. G. Wells, published in 1898. This literature persisted even after Antoniadi had denied the existence of the canals; it was based on another hypothesis: since this planet, so it was said, is farther from the Sun than the Earth, it must have cooled down faster. The assumption therefore was that it must be further evolved and that the "civilizations" of its inhabitants must be more advanced than ours. Wells already imagined that their planet was beginning to become uninhabitable: hence, the Martians were trying to invade ours. So they landed and began to ravage the Earth from their giant tripods. Due to an ironic quirk of history, it was Earthlings (in this instance the Americans) who in 1979 actually landed two "tripods" onto the soil of Mars from the Viking spacecraft. This was for entirely peaceful purposes, although the risk of contaminating the atmosphere of the red planet was taken quite seriously.

But when Wells wrote his book, the hypothesis of a Martian civilization had long had adherents among the best scientists. The great mathematician, Carl Friedrich Gauss (1777–1855) had already suggested that a huge equilateral triangle observable from Mars should be formed in Siberia by sawing down fir trees; the idea of lighting immense fires in the desert had been put forward for the same purpose. For a similar reason, the invention of radio immediately incited researchers to try to pick up signals from Mars. These attempts were repeated for several decades. About fifty years ago, when the Earth came close to Mars, an American astronomer asked that all radio transmissions be suspended so as to eliminate all possible interference and listen to

Mars in total silence. This was actually done, but no signals were detected.

The idea that "Martians" existed had totally lost credit among astronomers when it was decided to launch the Viking spacecraft (1976–1982). Its mission was to detect possible forms of primitive life. There was no plan to photograph moving beings, but to study soil samples picked up on the planet to determine whether they gave off carbon dioxide, which would have been typical of such life forms. It was very difficult to interpret the results of the three experiments performed, although they had been prepared with great care. These results were unexpected; after laboratory studies and simulations, the conclusion was reached that the planet is very rich in oxygenated compounds but that there is no primitive biological life in the first few inches below the surface, at least at the two sites that had been dug. It must be recognized, however, that the basic question has still not been answered. Some astronomers believe that life could exist elsewhere, deep underground, in the form of simple cells. So it was that exploration continued with the Soviet Phobos expedition (1988–1989) in particular, followed by the huge Mars 94 program for which the CNES [the French National Space Exploration Center] is to receive money and during which the Soviets will deploy in their atmosphere an explorer balloon developed by Jacques Blamont, director of the Aeronomics Laboratory in Verrières-le-Buisson.

All that can be said today is that there are apparently no multicellular beings either on Mars or on any other celestial body in the solar system, or even in interplanetary space.

Accordingly, research has rather been directed to the discovery of prebiotic life forms: either organic molecules that could be of great interest to biologists or prebiotic reactions that could prefigure the formation of complex molecules of the DNA type. This is why, as we shall see, interest is turning to Titan, a satellite of Saturn, to the comets, and to interplanetary space.

Some are also looking outside the solar system, at the nearest stars. But one can imagine that this investigation may encounter serious difficulties, considering that the closest star is four light-years away. One would have to travel a hundred light-years to find only a scant thousand.

How is such exploration to be accomplished? Today we do not have the technology to send people or spacecraft within proximity of these stars, even the closest ones. Although this year we managed to explore the solar system—one space probe (Voyager) has traveled twelve years covering over five billion kilometers [three billion miles], swinging near Neptune—we still do not have the means to reach the stars.

On the other hand, with electromagnetic waves we can obtain valuable information on what is going on in the depths of the Universe. It might be said that we can thus "observe" the Universe, but we must be clear as to the meaning of this word: we are talking about detection of radiation using increasingly sophisticated instruments. As far as electromagnetic waves are concerned, for example, the entire spectrum has now been covered: short waves (ultraviolet, x-rays, and very high-energy gamma rays) and long waves (infrared and very long radio waves). Our instruments thus afford us extremely far-reaching "observations." The illusion of the "celestial sphere," which has captivated humankind for millennia, is only a distant memory for today's scientist. True, to our eyes, the stars are still more or less luminous dots on a huge and sublime vault overarching the Earth, and they all look as though they are the same distance away. But we know that, in fact, there are astonishing differences in how far into space the stars range. Let us consider just three of them, visible with the naked eye in a star-studded sky on a fine night in the countryside: the Moon, the North Star, and the galaxy of Andromeda. Light reaches us at three hundred thousand kilometers [a hundred and eighty-six thousand miles] per second in a time frame of a second and a quarter from the Moon, in six hundred years from the North Star, and in two million years from the Andromeda Galaxy, a pale fuzzy spot at the limit of our vision! Our new means of observation have opened up unheard-of horizons. The most distant objects we perceive are "quasars," a hundred to a thousand times brighter than a galaxy; revealed by radio astronomy in the 1960s, these powerful radio emission sources look like stars in photographs (hence the name "quasar," an abbreviation of "quasi-stellar"). It was immediately apparent that these objects were traveling away from us at enormous speeds, entirely unknown for stars, on the order of one hundred thousand kilometers [sixty-three thousand miles] per second, while the stars in our galaxy travel at a few hundred miles per second at most. Hence, the quasars are racing away from us in a prodigious flight that suggests unimaginable distances. The "closest" of them, 3 C-273 (such an official and prosaic designation for such a wondrous object!) is three billion light-years away.

The expression "depths of space" is not an empty phrase; especially as such observations allow us, by the very nature of the case, to look back immensely far into the past, as a sort of premium. A quasar ten billion light-years away, for example, is seen as it appeared ten billion years ago, since its light has taken ten billion years to reach us.

But if the Universe really has these fabulous dimensions in space and time, an idea that dates from the past century, namely that of detecting artificial extraterrestrial signals transmitted by beings assumed to be more developed than ourselves, again becomes very current. It is based on three hypotheses which match present-day scientific data.

The first consists in assuming that life as we know it on Earth is the result of a natural evolution of the physical processes of the cosmos. Life no longer appears today as a principle different from matter. Since the "big bang" fifteen billion years ago until today, life in its extraordinary richness and immense profusion may be seen as the product of a grandiose evolution of the Universe. This idea has now taken hold with most researchers, but not without having to overcome intellectual and emotional resistance due in particular to the grip of theological doctrines.

The second hypothesis is to assume that what has happened on Earth over the course of four and a half billion years could have happened elsewhere in the Universe, in view of its immensity and age. If there are billions of stars in billions of galaxies, if the Universe is fifteen billion years old—that is, three times as old as the Earth—this hypothesis, like it or not, appears to be the most reasonable.

The third hypothesis is perhaps the most difficult to accept: human intelligence, of which we are so proud, does not represent the ultimate product of evolution. This may wound our narcissistic pride in being human, imbued as we are with our own superiority, but this hypothesis is perfectly reasonable if we just think about it, without bias. How can we not assume that during these billions of years in these billions of galaxies with these billions and billions of stars, evolutionary processes could not have led to more advanced results than those that happened on Earth? To put it another way, when we consider the seething evolution the Earth has witnessed in less than five billion years, if we think only of the last laps of the cosmic marathon that have brought "us" from Australopithecus to the Apollo astronauts, how could we believe that the 20th century on planet Earth represents the acme of the long history of the immense cosmos?

These three hypotheses seem to me *a priori* unarguable. Yet many have refuted them. This overt or veiled opposition, which may be betrayed by the merest smile or is often reflected in budget allocations for funding the type of research I shall be talking about, certainly has its roots in the subconscious: the same roots that for centuries denied intelligence to animals or even, let us not forget, to women.

Despite this profound resistance, work has been under way for over thirty years now in the United States by two physicists at Cornell University, Giuseppe Cocconi and Philip Morrison, since 1959. Those were the early days of radio astronomy. These two researchers had the idea of finding out how far the radio waves we were transmitting could be detected in interstellar space. They were able to show that, although this space was not as empty as had been believed, being sprinkled with free electrons causing interferences, the waves that propagated best were the "decimeter waves" (having a wavelength of about one decimeter) and that there was in the Universe a "natural" wavelength: that of the atoms of hydrogen, its most abundant chemical element. This wavelength is 21 centimeters, and it is unique and most remarkable. Our two researchers did not stop there: they suggested that an attempt could be made to detect possible radio signals coming from the closest stars that were clearly artificial.

At the same time, a student named Frank Drake, who was writing his thesis at the National Radio Astronomy Observatory in Green Bank, had the idea of adapting an existing receiver to see whether it would pick up radio waves at frequencies close to 21 centimeters coming from the nearest stars that most closely resembled the Sun.

This was the first experimental attempt to detect extraterrestrial radio signals of artificial origin. The first star yielded nothing; but the second produced such a spectacular result that Drake refused to believe it: "It was too simple to be true!" A maxim of utmost epistemological caution. After checking, he discovered that the signals he had picked up came from U2 aircraft, upper atmosphere military observation planes that were secret at the time but suddenly became notorious when one of them crashed over the Urals. These aircraft flew at an altitude of twenty thousand meters [65,000 feet]. So: one attempt and one false alarm. Since that time, new tests have been patiently repeated without yielding any conclusive results: 150,000 hours of listening have elapsed, with equipment trained on the two or three hundred closest stars resembling the Sun. Alerts have been received, but all have remained unexplained; there have been clear signals for which no natural explanation could be given, but these were not repeated. Thus, from a scientific standpoint the question is still entirely open as we cannot be satisfied with an isolated, unrepeatable, or unrepeated case.

One of the major reasons for the paucity of the results obtained is the low capacity of today's radio astronomy receivers. Thirty years ago, a receiver of this kind could receive only one channel at a time. What is a "channel"? Think of the human eye looking at a country scene. It is sensitive to various wavelengths, from red to yellow to violet. These electromagnetic waves are very different, ranging from 0.4 microns for violet to 0.7 microns for red. But, although the eye receives all these waves, even a sophisticated radio receiver of that day received only one specific wavelength. Drake was working on a single channel, the only frequency band to which his receiver was sensitive. Since that time, many improvements have been made to the equipment. Only ten years later, about a hundred channels were in operation. Today, in the best observatories of the world, radio astronomy receivers can tune in to a thousand channels simultaneously. These channels, for the reasons stated, are in the 21-centimeter range as well as another rather special wavelength in the cosmos which is derived from the OH radical. H_2O, water, is in fact an abundant element in the cosmos, but the molecule is generally split by ultraviolet radiation. So the OH radical remains an incomplete molecule and emits wavelengths close to 18 centimeters.

Monitoring has been done in this range, but not without serious difficulties because interference problems like those encountered by Drake are even more acute today: the communications, surveillance, and navigational satellites now cluttering up the Earth's atmosphere are a considerable interference with this type of research, with as indeed ordinary radio astronomy studying galaxies or comets. One might almost talk of veritable space pollution. For military reasons, North America has deployed a worldwide space surveillance system composed of telescopes and super radars which track all spacecraft, from the time they are launched to the time they eventually re-enter our atmosphere. Every object over 10 centimeters in diameter has been catalogued. Today there are over 7,000 of them, to which we must add some 50,000 fragments of debris larger than a bolt and 10 million larger than a shot pellet!

Be that as it may, it can be said that the "opening up" of a thousand channels represents considerable progress. However, there are no fewer than a hundred

billion possible communication channels in the wavelengths favorable for extraterrestrial listening.

The disproportion is obvious and dismaying.
This is why, as we shall see, NASA decided to launch
into a new technological adventure to construct a new
type of receiver which will have ten million simultaneous channels, making it ten thousand times more
powerful than present-day receivers. It is hoped that it
will go into service on October 12, 1992. A symbolic
date: the 500th anniversary of the discovery of America by Christopher Columbus. The Americans dream
of discovering, in turn, the cosmic Americas five hundred years after they themselves were discovered.

This is the great, and some would say crazy,
ambition of the SETI (Search for Extraterrestrial Intelligence) program which represents an investment of a
hundred million dollars over ten years. Below, I will
give readers the information they need to take the technical and intellectual measure of this prodigious
enterprise. But before we join the promoters of the
project in concerning ourselves about the existence of
signals coming from planets outside the solar system,
we should take stock of the irrefutable data we already
have about the existence of life forms in the Universe.

II

BIOASTRONOMY

In 1982, the International Astronomical Union established a special committee on "bioastronomy," a new discipline responsible for exploring life in the Universe. The first objective this committee set itself was to determine whether planets existed around other stars outside the solar system. This question, so simple on the face of it, is obviously a decisive one. If there is a chance of finding life, and even advanced life, it is indeed reasonable to search for planets other than those we know already that may be similar to our own.

Much progress along these lines has in fact been made in recent years, thanks to advances in theory, the development of cutting-edge technologies, and a huge harvest of observations supplied by space exploration over these past two decades.

In theory, we now know how planets may come into existence. We must go back to the process of how stars were formed from the initial gaseous medium. Computer simulations suggest that a planetary system may also form in such a hypothetical case: with wisps of residual gas around the star condensing into planets. If we follow this line of thinking, it becomes very likely that planetary systems other than our own exist, since the Sun is a very ordinary kind of star of a very commonplace model and has planets.

How can we detect their existence? To understand the methods used, we must first consider the solar system as a prototype. It contains the Sun, the Earth, etc. as well as a very large planet, Jupiter, which revolves around the Sun in about twelve years. Because of its enormous mass (317 times that of the Earth!) compared to the others, it may be assumed in a simple and easily managed model that it is the only one that revolves around the Sun. By way of reaction, the Sun then revolves around their common center of gravity. But this center of gravity is not at the center, but almost at the edge of the Sun, so that the Sun revolves in space around a point near its surface. This gives it a slight oscillating movement which counterbalances the revolution of Jupiter. This Sun-Jupiter system itself moves in interstellar space around the center of our galaxy, which makes one revolution every 250 million years. Let us now imagine an observer on a neighboring star: he or she sees that the Sun, instead of describing a nice, round circle, oscillates slightly on both sides of this circle. If this observer were able to measure this oscillation, he or she would have evidence that there is a planet rotating around the Sun.

This is the oldest method used to detect planets revolving around other stars. For more than 40 years now, two United States observatories with excellent telescopes have been using photography in an attempt to measure the apparent oscillations of the nearest stars. But the photographic technology used is not precise enough to lead to indisputable conclusions.

With today's electronic equipment, better results may be expected. We are trying to free ourselves from the uncertainties of the photographic plate by replacing it with a very fine grid of black lines and using photomultipliers.

Another method appears to have already been successful. Considering oscillation, if we imagine ourselves in the orbital plane of Jupiter, we will see the Sun sometimes advancing and sometimes retreating. Using a spectrograph, it is possible to measure the rate at which a star advances or retreats. The difficulty arises from the fact that the effect induced by Jupiter on the Sun, a typical one, is on the order of 10 meters [3 feet] per second. Hence the speed of the stars would have to be measured with an accuracy of 10 meters per second if this method were to be used. This was unthinkable a mere ten years ago. However, today we have a new and very simple technology (it just needed to be thought of) which points the way. We are indebted for it to a young Canadian astronomer, Bruce Campbell. The position of the spectral lines emitted by the star are measured with great precision by comparing them to lines created in the laboratory by the very spectrograph located behind the telescope. Traditionally, we were dealing with two rays of light: one coming from the star and one coming from the reference system. Campbell had the idea of sending them along the same optical path. By thus eliminating numerous sources of error, he achieved an accuracy of 10 meters per second. Of course, although the principle is of disconcerting simplicity, the technical implementation was extremely delicate and took years of work to develop.

The fact remains that now we have half a dozen stars which appear to meet the stated criteria for detecting a planetary system. Note, however, that Jupiter takes twelve years to revolve about the Sun, so that it would take about this length of time to draw conclusions about the existence of any planets found revolving around these stars.

But even assuming that the response were positive, we would obviously have no certainty about what kind of life could have developed there in one form or another. We know for a fact, for example, that life as we know it on Earth does not exist on Jupiter. And if we bear in mind that the Earth is three hundred times lighter than Jupiter, we see that, to detect a planet of this size, effects three hundred times weaker than those we are beginning to tackle would have to be measured. Such measurements could only be conducted from space, in order to escape the interference of the Earth's atmosphere. Gigantic projects are on the drawing board, involving optical or infrared telescopes 16 meters [52 feet] in diameter placed in orbit with extraordinary accuracy.

Let us add to this arsenal still other methods which exploit equipment not designed for the purpose, like Hipparchus, whose initial mission was to measure the positions of a hundred thousand stars with an accuracy of one-thousandth of an angular second. We can see that, secondarily, it would have been able to detect stars with oscillations. Unfortunately, we know that because of minor technical difficulties it would be difficult for Hipparchus to accomplish even its nominal program.

Detection of Earth-type planets will thus remain an open question for several years to come. We will be reduced to extrapolating from the study of more massive planets the size of Jupiter, if indeed any are found, to lighter planets.

However, one event must be reported that is not part of a deliberate program this time, but an unexpected discovery that came about, as so often happens, by chance. The Iras satellite, built jointly by NASA and the Netherlands, had the mission of studying the sky by complete and precise scanning in the infrared. The instruments had to be calibrated on board. The choice for this calibration fell on *Beta Pictoris* (the second-brightest star in the constellation *Pictor*, the Painter's Easel), a well-known star which had been stable for centuries. Surprisingly, it was found to have higher infrared radiation than it should have had. It was in accounting for this oddity that the existence around *Beta Pictoris* of an equatorial disc of dust and gas seen in profile was discovered. With a diameter comparable to that of the solar system, this disc, whose mass is equivalent to that of Jupiter, rotates around the star and could be thought of as a protoplanetary disc. One important fact: the dust is several microns in diameter, contrary to interstellar dust which measures only about a tenth of a micron. So it is believed that these particles, thought to be made partly of light-colored ice and partly of dark-colored rock, are the debris of colliding comets. This is a powerful encouragement for those who defend the nebula theories of planetary formation! Since then, at least a dozen stars with similar discs have been discovered. A word of caution: these are interesting facts but not, strictly speaking, evidence that planetary systems exist. They are only indicators that such systems might be undergoing formation.

What will come of the "exotic" equipment that some people are devising today to advance this research? For example, gravitational lenses which

amplify the images of quasars located behind galaxies? Fascinating as these prospects may be, we must
admit that they remain highly theoretical.

III

COSMIC

EVOLUTION

AND LIFE

Suppose that the existence of other planetary systems is confirmed during the next few years, for example by observations from the Hubble space telescope which has recently been placed in orbit; the next task would be to find, among the planets thus discovered, those that would be hospitable enough for life more or less similar to that which appeared on Earth to develop.

The emergence of life is indeed the outcome of a whole series of events which marked the evolution of the cosmos. The Earth is often referred to as the "ocean planet" or the "blue planet." The first astronauts were entranced by the spectacle of this bluish globe. This simple and beautiful appearance harbors two secrets of the life that developed there. First, the presence of large quantities of liquid water on its surface. Now, the very existence of this water requires extremely specific physical conditions: a temperature between 0° and 100°C and a pressure of 1 atmosphere. But that is not all. This liquid water must have continued in this form without breakdown for billions of years.

Other planets in the solar system did not have this chance. Remember Mars which is thought to have had plenty of liquid water three billion years ago. Since that time, when immense rivers—some of them as large as a thousand Amazons combined—were flowing over its surface, Mars succumbed to a glacial catastrophe. If there is any hope of finding water on this planet, it would only be frozen water in its subsoil!

Then think of Venus: an outright furnace; a runaway greenhouse effect, attributable to the enormous amount of carbon dioxide trapped in its atmosphere, heated it up to a temperature of 450°C.

This brings us to the second component without which life would not have been possible: carbon, which opens up the fabulous possibilities of organic chemistry. Properly speaking, organic molecules are just molecules based on carbon chemistry. Their impressive variety is due to the very specific properties of carbon atoms. They are capable of forming enormous frameworks or basic structures; molecules containing hundreds and even hundreds of thousands of carbon atoms can combine into aggregates of innumerable hydrogens, oxygens, nitrogens, etc. These structures can form chains as in the hydrocarbon series: methane, ethane, propane... with 1, 2, 3 carbons and 4, 6, 8 hydrogens; or they may form rings such as benzyl, diphenyl, etc. or rings in chains, branched chains... up to the helix of DNA, the carrier of heredity. Now, a very precise sequence of events was required for the right quantity of carbon to be established and for the terrestrial globe to be ripe for the adventure of life.

But if we are to understand these amazing chances, we must take another look at the entire history of the cosmos as we see it today. The particular evolution of the Earth makes sense only from this truly dizzying perspective, this grandiose fresco that physicists have traced, drawing lessons from Einstein's general relativity, quantum mechanics, and the observations accumulated for about thirty years by astrophysicists. Everything began with the "big bang." It created the volume of the Universe in its first "inflationary" phase so called. One second after the big bang, the Universe was just a thick soup of protons, neutrons, electrons, and photons at a temperature of ten million degrees. Under such conditions, nothing else could exist. A quarter of an hour later, 25% of the nucleons (or proton-neutron doublets) had been converted into helium nuclei by intense thermonuclear reactions. If the Universe had not gone through this expansion phase of unimaginable intensity, the cosmos would have retracted, shriveled, and collapsed onto itself. Life would have had no chance to appear because it would not have had the time.

Three hundred thousand years later, an eternity on the time scale of the initial events, we see the now "tamer" electrons being captured by nuclei and thus forming the first hydrogen and helium atoms. Then after a long period of lethargy extending over a hundred million years, when nothing happened except that the Universe cooled down and became darker, these atoms gathered into spheres, the stars, where nuclear reactions were triggered, producing heat and chemical elements. Did the stars first form individually and then group themselves into galaxies? Or did the stars spin off from the galaxies, as subunits? The debate is still going on, and there is still no theoretical or observational certainty on this point. In any event, the cosmos now awakened, and carbon appeared as did oxygen, the future basis of water, and silicon, which was to become the essential component of the Earth's rocks.

When they formed by condensation, the stars left wisps of material trailing around them which, in turn, condensed into planets. Let us now turn to the planetary system we know best, our own. We know its structure and history well. It consists of a massive central star, the Sun, in which thermonuclear reactions give off energy to a cortege of planets revolving around it: some dozen large ones, thousands of small ones, and further still, billions of comets scattered in an immense cloud of dust and gas.

And all of this, through which light takes ten hours to pass, is governed by gravitation and floats in space. Still further off, much further, light-years away, shine the stars.

44

What of the Sun's history? We have learned, from knowledge dating back less than half a century, that it was formed four and a half billion years ago, when the Universe was already old, by condensation of a primordial nebula of gas and dust spread out in space. We have begun to decipher the fabulous history of the formation of the Sun, our star. It took one hundred million years for an extremely bright proto-star to appear in the center of a sort of dense cocoon of gas and dust in a very violent phase known as the T Tauri star. This active center then blew the cocoon into a "stellar wind" whirling at hundreds of miles a second, while the star, now subdued, began its nuclear life. A magnetic field was created by an internal dynamo effect, and the "photosphere" formed at the periphery with huge spots. The magnificent solar corona hurled out its first magnificent arabesques.

The remaining gas and dust then settled into a disk where they agglutinated into light-colored ice particles and dark-colored silicate particles. This led to the formation of asteroids and the nuclei of comets; in a hundred million years, the planets as they are known to us today were formed. And among them was the Earth. A globe melted by the energy of falling. Too hot, and not solid enough, it was unable to retain the light gases, hydrogen and helium; all that remained were rocks and metals; it cooled down; and three billion eight hundred million years ago the first granite rafts began to float on its surface before forming the Earth's crust. The melting elements volcanically spewed forth gases: methane, carbon dioxide, nitrogen, and water vapor which formed the primitive atmosphere of the Earth. Then as the temperature continued to drop, the water vapor condensed and a more-than-torrential rain loaded with sulfuric acid beat down upon the soil. It leached it and swept the carbon dioxide along down to the ocean floor, where it was deposited in the form of limestone.

Thus it was that the Earth got rid of this gas and escaped the catastrophe that ruined its neighbor, Venus. No glacial catastrophe, no devastating greenhouse effect, no runaway volcanoes; neither too near nor too far from its star, the Earth enjoyed a temperature that was neither too low nor too high to maintain liquid water on its surface: the oceans. A long period of calm in fragile equilibrium allowed life to start developing and continue until it produced the forms we are familiar with today.

It is to life, created gradually over billions of years by organisms that developed systems of photosynthesis, that the Earth owes the massive presence of oxygen which makes its atmosphere unique in the solar system. It is also because of this oxygen that the ocean planet is also the blue planet.

How did life appear on Earth? This is still an enigma. But it is a fact that rocks three billion five hundred million years old contain fossil organic molecules, suggesting the rapid appearance of flourishing biotic activity. Could such conditions, the fruits of such a history, have been duplicated elsewhere, in other planetary systems? Once again, there is nothing to indicate the contrary. And when we consider the immense number of such possible systems, it is indeed quite reasonable to think so.

In any event, one more major fact lends credence to this idea, and that is the recent and impressive discovery of organic molecules in the Universe.

IV

THE ORGANIC STAGE

OF LIFE

IN THE COSMOS

What can be called the "organic stage" of life in the Universe has been a focus of research for about twenty-five years. Progress in radio astronomy has distinguished eighty different kinds of molecules in interstellar space. Most of these molecules are organic and can have up to thirteen atoms. An extraordinary discovery! How was it made? A molecule consists of at least two atoms. Here we are dealing with a diatomic molecule shaped like a dumbbell whose two weights can vibrate, and which has a rotational movement. When a molecule passes from one state of movement to another, it emits radiation whose wavelength, highly precise and identifiable, is specific to it and can be calculated in advance by the laws of quantum physics.

Now it turns out that organic molecules emit mainly short waves or millimeter waves. For a long time, radio astronomy was unable to detect such waves. This is no longer the case, and one discovery quickly followed another. The most complicated of the molecules discovered, with thirteen atoms, is in fact a series of eleven carbon atoms with a nitrogen atom N at one end and a hydrogen atom H at the other. These molecules are in a free state in a space that is almost a vacuum: in interstellar space, there is one atom every cubic meter. Even in the laboratory we cannot create such a vacuum so well. Hence the question: How could a vacuum, which should be cold, contain that many complex molecules? Even from the biologist's standpoint this complexity may still seem paltry.

A model has been devised to account for this apparently paradoxical situation. When a star reaches the end of its life, it can explode; this is called a supernova. The great Magellanic Cloud was a good example in 1987. But it is also possible for it not to explode: for a long time, it ejects its external parts into the atmosphere, creating what is known as a stellar wind. Such a star, already highly evolved, is rich in heavy elements of all kinds: carbon, silicon, etc. Imagine that a small grain of silicate, for example, is ejected in this way. As it is leaving the area of the star, water (H_2O), ammonia (NH_3), and methane (CH_4) molecules become deposited on its surface in the form of ice; then the grain heads out into interstellar space. It becomes much cooler and no further chemical reactions can occur. During this voyage, which can last hundreds of millions of years, the little grain is subjected to the bombardment of ultraviolet rays coming from every star in our galaxy. These ultraviolet rays shatter the ice molecules:

H_2O is broken up into H and OH, CH_4 into H and CH_3, etc. As a result, chemical radicals are released. Because the temperature is so low, however, they remain inactive. Then the grain happens to pass near a star; it heats up again, and the chemical activity of the radicals, thus awakened, is released. They will react with one another, combine... .

This model remains speculative, but it has given rise to laboratory studies. The Dutchman Mayo Greenberg established that the radicals diffused at the surface of the silicate grain at the same time that they were broken up; this means they moved, so that when they recombined, instead of being present in pairs as they were at the beginning, they almalgamate—to produce CH_3OH, for example. The adventure then continues. When the grain leaves the star it had approached, it breaks up again and will then recombine in an even more complicated form. In the end, the little grain of silicate will be covered by a coat of organic matter.

Other lines of research point to a different process, for which we are indebted to the Frenchmen Alain Léger and Jean-Loup Puget. Infrared spectroscopy shows that quasi-molecules of carbon can gradually form in space by agglomeration. Carbon atoms arrange themselves into hexagonal sheets, similar to honeycombs. Polyaromatic hydrocarbons are formed when hydrogen atoms graft themselves along the edge. Very recent studies have supplemented this picture. They are the work of specialists in soot, namely, those hexagonal sheets which slide over one another as in the domes of some sports stadiums. It has thus been discovered in the laboratory that there is a molecule (C 60) composed of 60 carbon atoms, which is believed could exist in interstellar space.

Whatever may come of this still tentative new research, we see that interstellar space, composed essentially of gas and dust, may become enriched with organic molecules. In particular, it is thought that the comets in our solar system may contain some of these organic molecules of primitive interstellar origin. This is why there is a great deal of interest today in studying the composition of comet nuclei in this light. The nuclei of comets, small celestial bodies a few miles across, produce spectacular tails when their substance is vaporized as they approach the Sun. For a long time they were thought to be made of ice. In 1973, Eric Gérard, research director at the Meudon Observatory, detected for the first time the chemical radical OH, a fragment of the H_2O molecule, in the comet Kohoutek with the giant Nançay radiotelescope in Sologne.

After a flotilla of space probes flew out to meet Halley's Comet, the portrait of its nucleus began to come into focus: a fragile, light fluffy conglomerate in the shape of a peanut 15 kilometers [9 miles] long, covered with a very black crust, and containing valleys, hills, and a three-kilometer [two-mile] impact crater. When the dust on its surface is examined, we see that half of it has a composition similar to some of the most primitive carbonaceous meteorites. But here is the most exciting discovery: the Giotto probe actually discovered large numbers of H, C, N, and O atoms, and it appears that a third of the dust is very rich in atoms which serve as a substrate for organic molecules; the presence of polycyclic aromatic hydrocarbons or even polyoxymethylene is suspected.

It could be imagined that this strange body, the fossil of a long-gone past, was condensed in the very cold external regions of the protosolar nebula which, four billion five hundred million years ago, created our planetary system. Some have concluded from these observations that the comets falling into planetary atmospheres were the origin of life. Models show that Venus, for example, and Earth in particular were bombarded by rains of comets, which could have seeded their atmospheres with organic molecules. Calculations have established that if everything had been preserved, the surface of the Earth would be covered by a layer of organic molecules of cometary origin a hundred meters [three hundred feet] thick. This may indeed have contributed to the origin of life; but these views remain controversial and precarious.

Similarly, amino acids have been discovered in meteorites. There is even one, the Murchinson meteorite, in which 55 different amino acids have been identified. Eight of them are among the 20 which constitute the building blocks of proteins on Earth. Add to that the fact that amino acids are dextrorotatory (the rotational movement they produce in light is to the right, or clockwise) or levorotatory (counterclockwise). Now on Earth, we have only levorotatory amino acids. This has excited the curiosity of researchers: some imagine that competition was at the origin of life and believe that one form prevailed over another by chance, while others, such as Evry Schatzman, attribute this situation to weak interaction. Yakob Borisovich Zeldovitch, in his most recent article, noted that the effect of weak interaction was inadequate to account for this situation, adding in particular that fluctuation effects could have led to this surprising result.

The fact remains that we would like to know more about meteorites. The Japanese and Americans have made considerable efforts to discover virgin meteorites in the Antarctic ice in addition to the three thousand and more that have already been catalogued on the surface of the Earth; the total is now at six thousand, and a 65-kilogram [145-pound] meteorite was recently discovered. Moreover, the Frenchman Michel Maurette found micrometeorites in the mud at the bottom of glacial lakes in Greenland. Some of them are in perfect condition and have a fluffy structure comparable to what we imagine the nuclei of comets to have been. The organic molecules in these meteorites can actually be analyzed. Another strategy for capturing meteorites is by stratospheric plane or on board the Mir space station.

Perhaps the most fascinating fact to emerge from these studies is the following: not only do some of these meteorites come from the Moon, which can be readily explained in view of its proximity to the Earth, but it appears that others come from Mars! How to explain this? Imagine the following scenario: large meteorites fell on the planet Mars in the first billion years of that planet's existence and the giant holes still observed may attest to this. Pebbles would have been hurled out, some of them attaining escape velocity. Thus ejected, they are believed to have rotated in the solar system like planets and from time to time fallen onto the Earth.

We may ask how the provenance of such an object can be detected. It is not that complicated; when a meteorite arrives in the Earth's atmosphere it leaves behind a trail of heated and hence ionized gas which, like a wire, reflects radar waves. From this it is not too difficult to extrapolate its initial orbit and determine whether, for example, it could have come from the planet Mars or from a comet.

V

THE LESSONS

OF TITAN

A new stage, which we will call "prebiotic." Here the molecules are more complex and heavier, although they cannot reproduce; they are often called the "building blocks of life." Our planet can tell us nothing of the formation of these blocks on Earth because it happened during the first billion years of its existence; since this far-off time, geological processes and various kinds of bombardment and erosion have wiped away every trace of them. Astronomers can now make their contribution to this formerly hopeless research. Let us turn to Titan, a satellite of Saturn, and one of the largest satellites in the solar system. Voyager I and II came close to it after twelve years of a flight that began in 1977.

This voyage in itself is worth dwelling on for a moment. Several years before the launch, NASA astronomers had noted that the giant planets Jupiter, Saturn, Uranus and Neptune would be in a favorable configuration for a flyby in a single voyage by gravitational rebound. They calculated that such a configuration would not recur until 176 years later. So they asked the U.S. Congress to fund what they called the "Grand Tour" with a specially designed space probe. The budget restrictions of the early seventies made Congress unwilling to do so. Undeterred, the NASA technicians then began to put together a lighter instrument that would still allow them to seize the opportunity. Thus Voyager I and Voyager II were conceived, with the official objective (as far as Congress was concerned), of flying by Jupiter and Saturn but not Uranus and Neptune.

In order not to miss the alignment, the launch had to take place within a three-week "window." The Voyager probes were equipped with booster rocket engines of the type that caused the shuttle Challenger to explode. First launch: Voyager II. No problem; the probe was well within the window even though it reached escape velocity with only 3.6 seconds worth of fuel left. The difficulty arose with the second launch, which could not take place until 16 days later, outside the window. Thus, Voyager I inevitably missed Uranus and Neptune! Here it was the booster that failed. If the reverse had been the case and the booster had been mounted on Voyager II, neither Uranus nor Neptune could have been reached!

Finally, the program did proceed partly according to the wishes of the NASA engineers. The Jupiter flyby was accomplished at a precisely calculated distance. Firing was adjusted en route so that the probe "rebounded" in the direction of Saturn and then, by a further maneuver, changed course once again for Uranus and Neptune. The next chance to resume the operation will not come until the window opens again in 2153!

The Voyager spacecraft flew by not only the planet Saturn but its satellite Titan, which has been known for a long time. With its diameter of 5000 kilometers [30,000 miles], or half that of the Earth, it is surrounded by an atmosphere which is as dense as that of our planet, essentially composed of nitrogen, as would be the case on Earth if it were not for the mass of oxygen of biological origin mixed with it. This characteristic is important: Titan (and now Triton too) is the only body in the solar system known to have an atmosphere composed of nitrogen, like ours. Another very interesting piece of information: since it is so far from the Sun, its surface is very cold (about $-180°C$) and, at this temperature, life could not have developed as it did on Earth. So this satellite might be considered a model of a primitive Earth that had been put in the freezer!

The most exciting finding was when the instruments on board the Voyagers detected the types of molecules that could exist in this atmosphere that is several hundreds of miles deep, and so cloudy that it hides the ground. The photographs taken showed two aerosol layers at high altitude, then lower down, cloud layers impenetrable to the instruments. The spectrographs showed the presence of organic molecules, in particular quite complex hydrocarbons—not only methane, which telescopes had already detected from Earth, but also acetylene, for example, and others. Most important, hydrocyanic acid (HCN) was discovered. This is a relatively simple molecule consisting of three atoms, but we know that adenine or guanine, that is, the nitrogen bases which constitute the rungs of the DNA double helix, can be polymerized from it in the laboratory! So we can ask the question: did a prebiotic stage start on Titan?

How could these organic molecules have been created in a world so cold and so far from the Sun?

For the basis of the answer to this difficult question we must look to photons: solar ultraviolet photons reach the upper atmosphere of Titan. There they collide with the atoms they encounter, and shatter them. When CH_4 (methane) is thus split into CH_3 + H (hydrogen), the latter, a very light atom, can escape from the atmosphere; on the other hand, CH_3 and N, derived from a split nitrogen (N_2) atom, can recombine. We can see how successive reactions could have led to an HCN molecule. Eventually, heavier hydrocarbons agglomerate to produce aerosol layers, and ultimately agglomerates heavy enough descend and fall to the surface of the satellite.

A final aspect of this program must be emphasized, because it opens up extraordinary prospects. If the explanation we have just summarized is the correct one, we would have to assume that all methane on Titan would already have been destroyed. Yet a small proportion of it remains. Thus, we must hypothesize that on the surface (which has not yet been observed) there is a planet-wide ocean composed of liquid methane or ethane.

Taking into account the temperature on the ground, the amount of ultraviolet rays radiating from the Sun, and the quantity of methane which is continuously recycled there, we may even calculate the composition of this ocean: in a temperature range from 10 to 20°C, it would be an ocean of pure methane or pure ethane (C_2H_6).

Other questions then arise. We know, for example, that nitriles (RCN) have formed. But are they soluble in this ocean? It all depends on its composition. Those that were not soluble could have formed sedimentary deposits. We might also imagine that continents could have extended over the possible ocean, made not of silicate but of water ice.

As we see, Titan is decidedly a very interesting satellite. So interesting that an astronomer at the Meudon Observatory, Daniel Gautier, suggested a plan about ten years ago of paying another visit to Titan and parachuting down a probe which, as it fell, could perform organic chemical analyses and study the physical composition of its atmosphere. The ESA (European Space Agency) and NASA have now joined forces to work out a project on this basis. What started out under the name "Operation Cassini" after the astronomer who was the first director of the Paris Observatory and discovered the multiplicity of rings around Saturn, will go forward if Congress does not object. The spacecraft will be heavier and more sophisticated than Voyager, designed over twenty years earlier. For a number of years it will observe the Saturn system, rebounding gravitationally to each satellite to study it several times; in passing, the probe, to be called "Huygens" in honor of the classical work of the Dutch astronomer who discovered the rings of Saturn, will be released into Titan's atmosphere. If everything goes according to plan, the Cassini-Huygens spacecraft will be launched in 1996 and the Huygens parachute will be dropped on January 12, 2003, at a specific time on that day! The calculations to be made are so precise that they must include the effects of Einstein's theory of general relativity. Even now, we can admire this theoretical achievement as much as its technological performance. And if all succeeds, in two or three hours the 90 lbs. of instruments on board will unveil to us many of the mysteries of this missing link of prebiotic evolution. All these endeavors and investments are clearly justified by the hope of gaining access, on Titan, to the conditions of genesis of life on Earth and solving an incredible enigma.

Could other satellites be of equal interest in the future? Among the satellites of Jupiter—Io, Europa, Ganymede, and Callisto—to judge by the density, temperature, mass, and nature of the surface of the second one (Europa), we believe that it is a kind of tellurian planet, that is, a great mass of silicate with probably a liquid ocean covered by a crust of ice. If this ocean is composed of liquid water, there is the possibility of complex organic chemical reactions, because water is a good solvent. But how to explain the existence of such a liquid mass on Europa? By the violent "stirring" this planet undergoes due to the powerful tidal effects caused by Jupiter. Stirring causes heating; we already know that such a process is at the origin of the volcanism on Io: under the gravitational effects of Jupiter, this satellite is covered with sulfur-spitting volcanoes. A strange and singular phenomenon—it is the only active volcanicity known in the solar system outside the Earth and now Triton—which is of the greatest interest to volcanologists.

This, in outline, is the knowledge astronomers have acquired about the prebiotic stage. It attests to organic chemistry reactions leading to already complex molecules, a possible pathway to nitrate bases, the essential components of DNA. These are the hypotheses they formulate, taking over from geologists to assist biologists in unveiling the mysteries of life.

VI

THE PRIMITIVE

BIOLOGICAL STAGE

Life flourished on Earth three and a half billion years ago, when the Earth was already a billion years old. Witness to this are the most ancient fossils, those of stromatolites (from the Greek *stroma*: carpet) still found in the seas around Australia in heaps of successive layers of bacterial residues about 50 centimeters [20 inches] thick. Stromatolites are colonies of monocellular beings that live in warm, shallow seas. Their oldest fossils go back three billion eight hundred million years! Some researchers claim that even more ancient fossils, cocoids, exist but this is still the subject of discussion. Indeed, the further back we go into the past, the smaller and more elementary the objects that must be identified and the more uncertain the interpretations given to them.

Research on the origins of life is not, of course, the prerogative of astronomers. It is the biologists, such as Stanley Miller, André Brack, or François Raulin, and many others, who have made it their main work, in the laboratory. They study the origin of proteins, nucleotides, and membranes—the fundamental stages of living reality—but the traces of this reality on Earth have been irretrievably obliterated by geological processes. Astrophysics may in the future contribute to this work, especially since space exploration of the solar system has yielded information on the composition of the atmospheres of all the planets. Now we know the winds that blow on the surface of Mars, the chemical composition of its atmosphere, and the prevailing temperatures and pressures. We possess the same information about Venus and have just seen what happens on Jupiter and Saturn. Mercury is a planet that has no atmosphere. Now that the era of comparative planetology has dawned, we can attempt to extrapolate a model showing the composition of the Earth's atmosphere at its beginnings.

Here again attention is directed in particular to Mars. You will recall that according to Janssen the atmosphere of Mars contained water vapor. Its measurement was very difficult because the water vapor in our own atmosphere distorted spectroscopic observations. The Meudon Observatory was created in response to this concern: Paris was already too polluted for such measurements. The great telescope of the observatory, 83 centimeters [33 inches] in diameter, still the fourth largest in the world, was suitable for such observations.

The attempts to find water on Mars had been unfruitful. Audouin Dollfus, some thirty years ago, had himself undertaken a stratospheric flight in a sealed aluminum gondola lifted by a cluster of a hundred ordinary weather balloons inflated with hydrogen. He accomplished the feat of rising to 12,000 meters [40,000 feet], trying to see whether, free of the Earth's atmosphere, he could still detect moisture on the red planet. The attempt was not conclusive.

The two Viking probes that landed on Martian soil in 1976 did not, at the spots where they dug, find any trace of primitive biological processes; but they did reveal very surprising traces of flowing water, in particular the bed of a gigantic dry river, the Kasei Vallis.

There is thus every reason to believe that Mars had a clement past before the glacial catastrophe, with liquid water on its surface. Where did this water go? Most probably into a frozen subsoil. In fact, the detailed photographs sent back by the Viking probes revealed at least a thousand very special impact craters: under the shock of meteorites, each crater was wreathed in a lobar halo resembling petals of flowers lying on the ground—just as when a heavy stone is thrown into the encrusted dried-out bed of a muddy lake. These petals are probably due to ejection of a frozen subsoil, what is known as "permafrost"—a sort of concrete made of pebbles and ice—which liquefies under the shock and freezes as it emerges.

François Costard of the Meudon-Bellevue Physical Geography Laboratory investigated to see whether the permafrost could have cropped out at certain points, and found at the outlets of the gigantic ancient flows into the northern plains, Chryse and Utopia, that the ice could be as little as 60 meters [197 feet] deep. These spots will of course be favored sites for the upcoming Soviet expeditions in 1994. Perhaps then we will be able to answer two exciting questions: Since the craters in question date back several billion years, is the frozen water still there? Will we then discover a primitive biological life independent of ours, similar or different, fossilized or frozen, in the subsoil of Mars?

VII
EXTRATERRESTRIAL
INTELLIGENT
LIFE?

Let us change scale once again and pick up the history of the Earth where we left it when we described the grandiose cosmic evolution that led to the appearance of life. You will remember that the first continental rafts solidified and the oceans condensed. This happened three billion eight hundred million years ago, only a very short time in geological terms so to speak after the formation of the solar system. At that time, living organisms were monocellular creatures that had already diversified into several evolutionary lines. These organisms are very small, no bigger than a micron in size.

It took another billion years for photosynthesis to be invented. A marvelous and powerful invention, this is the process by which green plants, algae, and a few bacteria capture the energy of light and use it to synthesize organic compounds. Thus began the massive production of oxygen which would steer the atmosphere of our planet in a direction favorable to life.

This major stage was followed, about one billion four hundred thousand years ago, by the appearance of eukaryotic cells. These cells, a thousand times the volume of the existing bacteria, are already genuine complex factories with specialized shops: a nucleus for DNA, mitochondria for respiration, chloroplasts for photosynthesis, Golgi apparatuses for excretion, ribosomes for protein synthesis, and even flagella for movement. These cells thrived in the oceans for nearly a billion years. This biological activity increased the oxygen content of the oceans. Even more complex living forms appeared about 670 million years ago: the Edicara, the first multicellular beings, that would reign for 120 million years. Soft-bodied marine creatures, most of them looked like leaves. Being very flat, they presented a maximum of their surface in contact with the water from which they picked up oxygen, the necessary fuel for their evolved metabolism.

Then came what has been called the "Precambrian explosion." Five hundred and fifty million years ago, oxygen became abundant enough for an ozone layer to be created, which for the first time protected the ground from ultraviolet radiation. The continents were then colonized by life forms of an extravagant diversity, enough to make the best science fiction writers pale with jealousy. To cite only the animal forms: sponges, worms, anemones, insects, starfish, octopuses and vertebrates. The vertebrates, with many variations in their different classes, were fish, reptiles, birds and mammals; the mammals were carnivores, insectivores, ruminants, marsupials... and primates. And among the primates, anthropoids,

A dazzling profusion which took only a tenth of the age of the Earth to develop and for there to appear, among the gibbons, orangutans, gorillas, and chimpanzees: Australopithecus, the inventor of walking upright.

Intelligence then dawned in humans, if not exclusively at least in greater abundance than in other animals. Yesterday, *Homo sapiens* set foot on the Moon. I have two photographs that I like to compare: one of Australopithecus traces found in volcanic ash in Tanzania, dating back three and a half million years, and the other of the footprints of the first person to touch the soil of the Moon during the Apollo mission. When you look at these two photos together, you have, in summary, a complete view of the history of the march of the human species: from the ash of Tanzania to the dust of the Moon in three and a half million years. An extraordinary adventure, whose first photograph, that of the footprints of our far-off ancestor, reveals one of the major secrets. You see two sets of footprints: on the right, those of two adults, with the woman carefully placing hers in those of the male who walks in front of her, and next to them, on the left, the child whose strides are as long as those of the parents. Already he or she is trying to imitate them: a human trait to which no one can remain insensitive and which, with others, tells of our progress during our evolution. Lest we forget, however, this evolution has taken up altogether only one thousandth of the Earth's age!

Here, in broad outline, are the forms that life took until it evolved into its most complex form: human intelligence and consciousness. How can we fail to be struck by the fact that, on our tiny planet in the immense cosmos, this intelligence has emerged only in the last ten million years? Can we still believe that the entire evolution of the cosmos was directed to the creation of this intelligence as its consummate achievement? How can we not concede the possibility that other events of the same type, other evolutions, could have occurred far earlier elsewhere in the billions of galaxies that have populated the Universe? If this is the case, these evolutions would have produced more highly developed forms of this intelligence.

We must acknowledge that this prospect is a fascinating one: if we were to discover such intelligence, we would suddenly have a glimpse into our future and an answer to the questions that we have never stopped asking but have never had the wherewithal to answer. Let us remember once again that three and a half million years on the geological time scale is nothing. What will we become when another such period of time will have elapsed? It is totally impossible to come up with a reply by extrapolating from the present—as impossible as it would have been to predict the present status of civilization from the progress, already remarkable, of the Neolithic age. Even if we could look ahead a mere three hundred thousand years, we could not validly anticipate the future!

Just as astronomers can furnish the elements of a response to the question of the origin of life, where geologists must remain mute for lack of analyzable, interpretable traces, it is also not impossible that they can even today supply valuable information on the possible future of our species which could, moreover, help us to improve it before it becomes "our" present.

Such is the philosophical and practical meaning of the extraterrestrial intelligence research and listening program called SETI. But before we define its goals and describe its formidable resources in detail, I must answer one last objection of principle.

You are postulating, people may say, the existence of intelligence elsewhere in the Universe. You flatter yourself that you are breaking with the ageless anthropocentric prejudice of humankind. Fine. But must the intelligence of which you speak be conceived of as being the same type as ours, linked to the development of the singular central nervous system whose developmental laws we are just beginning to glimpse? Are you not falling into another trap in human thinking which is just as ageless: anthropomorphism?

My reply is that we are not at all prejudging the form of this intelligence. Witness the great astrophysicist Fred Hoyle who indulged in a number of speculations on this subject in his science fiction writing. The least that one can say is that he stays as far away from anthropomorphism as possible. His magnificent novel, *The Black Cloud*, written about thirty years ago, tells the story of magnetic interstellar clouds which are living beings: these clouds have magnetic tubes which channel electrons along specific lines of force....

But let us leave this kind of highly risky extrapolation right there! It is time to get back to SETI.

SETI is an acronym invented by Philip Morrison, a leading physicist at MIT in Cambridge, Massachusetts; together with Giuseppe Cocconi, a cosmic ray specialist, he published (in 1959) the first theoretical article on the possibility of communicating over interstellar distances using radio astronomy techniques. I had the good fortune to work with them at Cornell University, where I did my thesis.

At the same time, in 1960, Frank Drake, then a young student at the NRAO, the largest radio astronomy complex in the United States, had, on the basis of relatively simple calculations, proposed to his boss, Otto Struve, adapting a radio astronomy receiver to pick up signals on the wavelength recommended by the theoreticians (the 21 centimeters of the hydrogen atom). He wondered whether he would receive signals on this wavelength from the two closest stars most resembling the Sun: τ Ceti and ε Eridani. I have already mentioned the disappointing results of his observations: the very sharp signal he received was in fact attributable to U2 aircraft, which were secret at that time.

SETI lies within the scope of this line of research. We know that there is nothing corresponding to intelligent life in our solar system, so we have to look to the stars, beginning with the nearest ones. But we must understand what we mean by "nearness." While light takes a few hours to pass through the solar system, it takes four years to reach the nearest star. Even the closest stars are so far away that today it is unthinkable to obtain observational data using space probes that would "go have a look." Hence the selection of electromagnetic waves for detection. We could indeed think of other methods; for example, neutrino fluxes such as those emitted by the supernova of 1987 and picked up by a small number (a dozen or so) of detectors on Earth. But neutrinos are very hard to handle because of their ability to pass through any body they encounter.

So, electromagnetic waves are used, with the additional advantage that they are very fast: they move at the speed of light which, as we know, according to Einstein's theory, is the greatest speed that can exist in the Universe.

But very precise wavelengths must be selected. At long wavelengths, just as when an optical telescope is pointed at the sky in broad daylight, we have some difficulty in observing the stars, as the background of the sky is very intense. It is so intense because the stars in our galaxy emit an enormous number of natural radio waves at wavelengths greater than one meter. This situation is attributable to supernova residues, ionized interstellar gases, electrons traveling in the magnetic fields of our galaxy and radiating according to the synchroton effect (just as in particle accelerators). At short wavelengths, on the other hand, observation is blocked by the quantum nature of light. Electromagnetic waves are waves and corpuscles at the same time. The shorter the wavelength, the greater the intrinsic energy of each associated corpuscle (in the case of visible light, for example, a photon). Let us take an x-ray photon, which has enormously more energy than a visible light photon. If we sought to engage in interstellar communications and chose very short wavelengths, these would lose their wave nature and become bursts of corpuscles. One consequence would be that for a given energy budget in a given frequency band, the number of particles that could be received, say each second, would be fairly limited. To send information selecting a wavelength that would give only a limited number of "bits" would be impossible. Result: wavelengths shorter than one centimeter are not propitious.

If we take into account all these difficulties, not to mention impossibilities, without even considering the disadvantages of atmospheric phenomena, from which we may be freed in a few years by equipment actually placed in orbit around the Earth's atmosphere, we conclude that the electromagnetic waves most suitable for transmitting information have wavelengths between 3 and 30 centimeters; that is, radio waves corresponding to frequencies of 1 to 10 gigahertz. We can form some idea of these frequencies if we look on the dial of an FM radio: the frequencies used are in the vicinity of 100 megahertz. Remember that in the case with which we are concerned, a frequency of 1 gigahertz, or 1000 megahertz, corresponds to a wavelength of 30 centimeters. The SETI program proposes exploring wavelengths between 3 and 30 centimeters. We can see that its ambition is philosophically defensible, theoretically justified, and technically feasible.

Let us add that for the time being we are not envisioning "conversations" with extraterrestrial beings: for a star a hundred light-years away, any signal it would send would take a hundred years to reach us, our answer would take a hundred years to reach it, and the response necessary for any kind of real conversation to take place would take another hundred years to come back. I may be forgiven for doubting whether enormous funds would be released by any government agency with a "benefit" prospect calculated in centuries! On the other hand, if a star emitting signals is detected, it is probable that humankind would decide to embark on a "conversation."

Now that we have covered the principles involved, let us take a look at the implementation of the program.

VIII

THE PROSPECTS

OF SETI

Technology has led our generation to take an important step in the exploration of the Universe. With radiotelescopes of the Arecibo type, with a diameter of three hundred meters [close to a thousand feet] and a radar capable of sending out very brief but very powerful pulses to the planet Venus to study its relief, we have taken the interstellar step. We are now able to listen to the stars. Fifty years ago, in the early days of radio, every effort was made to listen to the planet Mars. But the stars were hopelessly out of reach.

Yet, this cutting-edge technology, marvelously powerful and sensitive as it is, is still very limited in its performance if we consider the immensity of the space to be observed. We have explained why the waves most auspicious for communications over interstellar distances have wavelengths corresponding to frequencies of 1 to 10 gigahertz. From the standpoint of information theory, we may add that the sensitivity of communications will increase in inverse proportion to the width of the frequency band occupied by the waves that are used. Anyone can test this for himself: in very noisy traffic, for example, or the slightly more hushed uproar of a cocktail party that is about to end, if someone addresses you ten yards away in a fairly low voice with pitches ranging from deep to shrill, you will not hear him or her distinctly; but if in the same situation this person decides to blow a whistle, even softly, the energy it emits will propagate in a very narrow frequency band and you will understand the person very clearly. Generally speaking, for transmitting information in the most effective way, the waves carrying it must be located within a very narrow frequency band.

But the very special and most formidable difficulty when we are dealing with stars arises from the fact that these waves must pass all the way through interstellar space before they reach us (or reach them, if we are transmitting). Now, even though this space is practically empty, it does contain electrons which will affect propagation somewhat. If a wave (or a radio photon, which comes to the same thing) strikes such an electron, it will be modified. When this wave reaches us, both its energy and its frequency will have changed. Hence we conclude that if extraterrestrial beings were to send out a beam of very specific waves, they will be received over a far wider band. Theory shows that there is no point in transmitting in bands narrower than those corresponding approximately to one tenth of a hertz. If we take into account the "window" to which I have already referred—the SETI window from 1 to 10 gigahertz—, we can calculate that there are a hundred billion possible channels in this window, each representing a band of one-tenth of a hertz, available for interstellar communications! Remember that our pioneer, Frank Drake, had only one channel available; at the present time, radio astronomy receivers have only a thousand simultaneous listening channels. But despite spectacular progress, the gap between a thousand and a hundred billion is an enormous one. It was to remedy this situation that NASA decided to build a receiver ten thousand times more powerful than those in service today.

The history of this project is not lacking in interest for those who wonder about the social realities of present-day research and the complex relationships between science and government. It all began at NASA's Ames Research Center in California, which works with another NASA institution, the Jet Propulsion Laboratory. These two organizations work with the Radiotelecommunications Laboratory of Stanford University in California. For the past ten years, this laboratory has been working toward a new type of receiver using technologies that have emerged from computer science: the waves received are processed by sophisticated mathematical methods. These methods, which may be termed revolutionary, will prove to be very useful when the waves received are to be cut up in a certain bandwidth into ten million segments, as NASA hopes.

Of course, NASA's activities are not limited to the SETI program: far from it! After the beginnings which I have just sketched, a freeze on the funding necessary for SETI research was imposed until 1982, due to an "ideological" veto by one member of the Senate. Since then, the initial SETI Project Office has become a full-fledged department within NASA, on the same level as the Solar Science Department, for example. This is an indication of its importance today in the eyes of the NASA executives.

Now that SETI is no longer a project but a highly structured program, what does it look like? The essential technical feat will consist of cutting up the radio bands, according to mathematical operations called "Fourier transforms"—named after their inventor at the end of the 18th century, who, for example, converted any noise into pure sound—into one hundred segments which, using the same principle, will in turn be cut up into one hundred new segments. This will give ten thousand segments which will be cut up into one thousand, thus yielding the desired ten million channels! The technology used is extremely elaborate; it is based on microchip technology. The prototypes already made have shown the effectiveness and accuracy of the method.

But it is not enough to cut up radio bands so finely. We also have to observe the "composition" of the segments. If I may use a trivial but telling analogy, let us take a salami and say that what we are interested in is not just cutting it up into superthin slices, but detecting the presence and alignment of any pepper grains in them. So it is necessary to observe ten million slices to analyze them. This is a veritable *tour de force*! Take a star: observation will last a thousand seconds, giving rise to a stack of a thousand observations, each lasting one second. If we want to examine the results of ten million slices of the radio band transmitted, we will have to accumulate a thousand times ten million slices, spreading them out to see which have "grains of pepper." Then we have to find out whether, for example, fifty grains are in a regularly spaced group. The operation can only be accomplished by supercomputers developed specially for this program. These supercomputers are specialized in what is called "shape recognition"; they are particularly suitable for detecting artificial signals. Such instruments are now familiar to us in, of course, cruder versions. Think, for example, of the robots which pick up parts in Renault's automobile factories. They must recognize the part, hence they must be able to identify its shape, but this is not enough: the robot must be able to detect the position of the part, pick it up from the right angle, and fit it together with the part it goes with, etc.

But in the case of the SETI observations an additional difficulty arises, and not a slight one at that: unlike the case of the automobile assembly line, here the shape is not known in advance. The aim is to recognize whether there is a shape that could be interpreted as artificial. Because pragmatically the first step has to be the simplest one, the first goal will be to try to detect two shapes which were considered *a priori* to be characteristic of such artificiality: a regularly spaced beep-beep or a continuous signal. Once this admirable receiver becomes operational, new programs will of course be written for recognizing more complex shapes. It is also planned to install more powerful detectors and "slicers" which will be able to bring the number to one hundred million and then to a billion.

The present "slicer" is called the MCSA (Multichannel Spectrum Analyzer). It is based on a cascade of mathematical operations performed by a computer of modular architecture with hundreds of microprocessors in parallel assembled around VLSI (Very Large Scale Integration) chips designed especially for the purpose. The MCSA is accompanied by a signal analyzer whose job it is to recognize in real time the existence of any regularities that could betray their artificial origin. We can get an idea of the enormous task to be accomplished if we know that a typical observation of a thousand lines each containing ten million points, with each point having an intensity coded from 1 to 100, represents a trillion values; within these trillion values, like snow in a picture, we must pick out a dozen or so flakes with a regular alignment; and this must be done before the next observation, as there is no question of storing the data, which is mainly noise.

The prototype for the MCSA was based on the technology of printed circuit boards in electronics. As we know, these boards are very small, 10 to 20 centimeters [four to eight inches] in size. Nonetheless, even if they were not reduced still further, the final SETI receiver would be gigantic (150 racks, or the equivalent of 150 average-size sets of bookshelves). This would take up too much room and consume too much energy. Today, therefore, we are going on to the next stage with a view to the final model. This stage is that of the even more compact "supermicrochip." A 10- to 20-centimeter [4-8"] board filled with chips will be replaced by a single microchip. It is also worth noting that these supermicrochips are themselves developed by computers. So today's computers are making the computers of tomorrow!

These electronic instruments will of course be mounted behind a radiotelescope, as large a radiotelescope as possible. The largest is the one in Arecibo. But it is a university telescope belonging to Cornell; this has many advantages but also, in the case that concerns us, some annoying drawbacks. It is open to all scientists, not only Americans but scientists from the entire world. All you have to do to be able to use it—not an easy matter, however, because competition is very keen—is to present a solid, precise program of observations. This program is submitted to a committee which allocates observation time. Because of these constraints, NASA would be alloted only one week of SETI time every three months, which is obviously inadequate, the more so as the field of observation at the telescope is limited to a portion of the sky in the vicinity of the zenith in Puerto Rico, where it is located.

Of course NASA has other resources: it owns radiotelescopes in its Deep Space Network, comprising the bank of instruments for remote control and monitoring of deep space probes like the one that left the solar system in 1989 after passing near Neptune. But even the largest radiotelescope in this huge network is only 70 meters [230 feet] in diameter, which is not much.

Happily, there is another large radiotelescope for decimeter waves; it is in Nançay in the Sologne (north-central) region of France. It is a rectangular telescope, 300 meters long by 40 wide [984 by 131 feet]. This is only a tenth of the area of the Arecibo telescope, but it is the second-ranking telescope in the world for the wavelengths that are of interest to SETI. The decision to invest in this radiotelescope was a bold but, as it turned out, a very wise one; we owe it to the pugnacity and clear thinking of Yves Rocard, working twenty-five years ago with Jean-François Denisse. Thus France has a highly valuable resource that may enable it to participate actively in SETI, even though its directors have not yet seen fit to allot a significant budget allocation to this program.

So in 1985 I launched the idea of an original "exchange": when NASA's reception and analysis instruments are completed in 1990, it would make a duplicate that could be set up behind the Nançay radiotelescope; in exchange NASA would receive observation privileges of, for example, one month a year, which would be very useful to it in view of the situation I have just described. This MegaSETI project, which is the term we have begun to use, could also be run on other instruments, such as the planetary telecommunications antenna being planned by the CNES (the French National Space Research Center).

Both the Nançay and the Arecibo radiotelescopes are already twenty years old. Technological progress has been such that they are now nearly antiquated. So it has been decided to modernize the Arecibo focal system by installing a so-called "Gregorian" double-mirror focus, an improvement that will benefit the entire community of radio astronomers by creating a new-generation instrument. In 1989 I succeeded in convincing the person responsible for this modernization to take a look at the Nançay telescope, which is very similar. A preliminary study is under way, which demands reams of calculation on highly powerful computers. If, as I hope, renovation of the Nançay focusing system is accomplished, our radio telescope will become four times more efficient than it is now. Not only will this represent an additional asset for SETI, but it will also be of considerable benefit to researchers observing galaxies. If an observation that now takes an hour can be completed in a quarter of an hour, it will soon be possible to observe four times as many galaxies from the Earth's surface, a virtually revolutionary development.

This is one of the "spin-offs" of SETI which, with other projects, should encourage governments to authorize funding! For its part, NASA is not hesitating.

How far along are we with the program? In recent years, complete prototypes with a hundred thousand channels have already operated, as well as MSCA with two million channels and a simplified analyzer. Two types of programs are planned to last half a dozen years: targets and scanning.

Research on targets will focus on the eight hundred stars most similar to the Sun up to a distance of a hundred light-years; it will operate at maximum sensitivity with channels of one hertz. This research is designed to detect signals from civilizations that might have evolved under planetary conditions comparable to ours.

Scanning research will systematically scan the entire sky but at lower sensitivity, with channels ranging from 32 to 1024 hertz. Here the goal is to pick up any powerful radio beacons, whether intentional or not, of unknown nature and not necessarily planetary.

I do not want to end this picture of the SETI program without describing one very concrete future prospect that I am engaged in working out with colleagues at the Meudon and Nançay observatories, including François Biraud. It will then be time to ask about the general philosophical implications of this fine-tuned research.

The NASA program, powerful and extensive as it may appear, calls for the observation of only eight hundred stars as we have just seen, during the six years following the startup of the new receiver. Within the context of such a program, each observation will only last, in fact, thirty seconds per star at a given frequency. Could not the efficiency of these magnificent instruments that are coming on line be improved still further? It was with this in mind that I proposed a listening strategy based on pulsars. A pulsar is a very small celestial body 10 kilometers [6 miles] in diameter which comes into being when a normal star collapses upon itself at the end of its evolution; it is what is called a neutron star, which revolves on its own axis at the dizzying speed of up to hundreds of revolutions per second, and whose magnetic fields are enormous. These magnetic fields generate radio beams which, whenever they sweep the Earth, give us a radio pulse. Thus, pulsars are natural powerful interstellar radio lighthouses, well distributed throughout our galaxy. Their pulses, due to their rotation, have periods of impressive regularity and their life span exceeds a million years. They are clocks of unequaled precision. Why shouldn't they be taken as common references by civilizations engaged in intentional interstellar communication?

First we have a problem of choice. Which reference pulsars should be selected under the SETI program? Everything depends on the type of research in which one is engaged. In the case of target research, for the eight hundred closest stars similar to the Sun, the nearest pulsar should be used. If we are researching a cluster of stars we take the most powerful pulsar. Then for scanning research, we will select the closest pulsar in the direction being explored at any particular time.

Finally, there is one difficulty that remains to be surmounted if this strategy is to become operational. This difficulty stems from the fact that the rotational frequency of the pulsar is generally very low (1 to 1000 hertz). This is far outside the frequencies that fall into the SETI window. Therefore a means had to be found for converting these low frequencies so that the result would come within the window. I proposed a very simple mathematical procedure which consists in multiplying the frequency of rotation of the pulsar selected as many times as necessary by a particular universal mathematical number (the number "π", for example) under ten (ten being the relative width of the SETI window) until it does fall into the SETI window. It is by such frequencies based on pulsars that I propose to begin exploring the galactic window to increase the possibility of early success, and I undertook to experiment with this listening strategy with my colleagues at Meudon and the SETI Project Office.

These days, SETI is a program supported by prestigious international organizations: the Academies of Sciences of the United States and the Soviet Union, the International Academy of Astronautics, and NASA. The activities of SETI are deployed throughout the world in the form of colloquiums, meetings, and various forms of cooperation. It may indeed be hoped that we will make the historically symbolic date of October 12, 1992.

But the picture of the search for intelligent life in the Universe would be incomplete if I did not describe another approach, different from that of SETI: the Soviet approach. The Russians themselves launched a program that they intend to develop in the years to come, after fifteen years of sleep and stagnation.

The same philosophy inspires this program, a philosophy which resolutely turns its back on anthropocentrism. It was explicitly with this perspective that a young Soviet student, Nikolai Kardashev, in the late 1950s, forged the idea of several types of civilizations. He distinguished three types. A Type I civilization appears as one capable of manipulating its environment on its planet-wide scale: we are nearly there, since human activities are very substantially changing the face of the Earth and since the question arises as to which decisions should be made for such "manipulation" to be controlled and beneficial, without precipitating a disaster.

A Type II civilization would be one capable of manipulating its central star. We use the energy that comes to us from the Sun in all kinds of ways, but between this energy and that of the Sun itself there is an immense factor—billions and billions. Such a civilization would thus have a power billions of times greater than ours.

A Type III civilization would be capable of manipulating its entire galaxy. There we go to a scale a hundred billion times greater still.

Based on these daring views, Kardashev came to imagine the possibilities of transmitting considerable volumes of information over interstellar distances. For example, an encyclopedia could be transmitted in ten seconds! This implies enormous transmission energies and frequencies occupying very broad bands. Thus, by radioastrophysical exploration, he hit upon quasars without being aware of it. A brilliant proof, if one were needed, that speculation, even apparently unbridled, can still today lead scientists to major discoveries.

As far as we are concerned, these general views imply a very different listening strategy from the one which developed in the wake of Cocconi, Morrison, and Drake. Attempting to detect a "Type II" civilization, the Soviets centered their research on special waves: millimeter waves. And in Samarkand they are now building the largest radiotelescope in the world for such waves. Kardashev, who is in charge of its construction, starts from the idea that a Type II civilization, even if it is not addressing an intentional message to the other stars, necessarily produces substantial and detectable technological leaks.

He can use the fact that, unintentionally, we on Earth, in our Type I civilization, are already transmitting along the lines of the case he is picturing. Radio, as we know, has existed for eighty years. And so, waves broadcast in these early days have already reached points eighty light-years away from us! They are also touching hundreds of stars of the Sun type. You might say these old waves were weak. Probably so, but who knows whether the "receivers," if any, have extraordinarily sensitive instruments?

In any event, because of the very powerful military radars we have today, we are reaching a million stars; even if their signals have not yet arrived, we know that between ten and twenty years from now they will have traveled ten or twenty light-years.

Think for a moment. "Those" picking up our transmissions, what could they know about us? Take one of the improved receivers like the great radio telescope in Arecibo, Puerto Rico, and set it up on the star Proxima Centauri four light-years away. Such receivers could easily pick up the carrier waves of our television transmitters! We know that such a transmitter produces a beam of waves traveling along the ground, but a good portion of these waves escapes into space. Most of the energy transmitted is contained in what is called the "carrier wave," a sinusoidal wave of a very precise and typical frequency. To this must be added the modulations which provide the picture and the sound.

Thus, from Proxima Centauri one would see three times every twenty-four hours a burst of broadcasts corresponding to the three intensive television areas: Western Europe, the United States and Japan. The "beings" on Proxima Centauri would thus be sure that these signals were artificial. They could easily conclude that they come from an object which rotates every twenty-four hours. Then, thanks to the Doppler effect (shift to the "red" of waves transmitted by an object that is moving away), by very precisely measuring the frequencies of the carrier waves, they would determine the speed at which the

Earth rotates on its own axis (300 meters per second or 980 feet per second). By combining these two bits of information (speed of rotation and periodicity of twenty-four hours), they would deduce the diameter of the object, that is, the Earth, as being roughly 10,000 kilometers [6,200 miles]. Assume that, knowing the diameter of the Earth, they continued observing and discovered an additional Doppler effect derived from the fact that the Earth is revolving around the Sun, they would determine a speed of approximately 30 kilometers per second [19 miles per hour]. This star, the Sun, would be thoroughly familiar to them through spectral studies of the same type as those that enable us to know of the existence of Proxima Centauri. Since they have determined the mass of the Sun, it would be easy for them, knowing this mass and the rate of revolution of the Earth, to deduce the distance between the two bodies: 150 million kilometers [93 million miles].

Since the Sun emits energy, they can also deduce how much of it the Earth, a globe 10,000 kilometers [6,200 miles] in diameter, located a 150 million kilometers [93 million miles] from it, receives. They would deduce that this quantity is such that liquid water could exist on its surface. Hence the conclusion: A civilization based on liquid water could have developed there.

Better still, the beams emitted by television transmitters travel along the ground and are absorbed to a greater or lesser degree according to the state of the vegetation, depending on whether or not there are leaves on the trees. They could thus deduce that there are seasons on the surface of the Earth. They could even draw "political" conclusions from their observations, noting for example that, until today, the broadcasting schedules are not the same in the USSR and the eastern bloc as in the western countries. If they were very intelligent, they would see that, in the final analysis, this difference reflects different economic fundamentals and different lifestyles, hence unequal development.

Pure speculation? Not really; remarkable studies have been conducted at Green Bank, Virginia on the reflection of our waves on the Moon! From these reflection effects, it has been possible to deduce extremely precise positions of our military radar, and to determine their longitudes and latitudes. Exciting, extraordinarily telling, and... confirmed results.

We see it. Even if we do not receive coded signals intended to be interpreted by the inhabitants of the Earth or any other planet, there could exist in the Universe technological leaks coming from other systems, similar to those from our civilization. This is the point of departure of the Soviets who are researching broadcasts transmitting a great deal of information in these far broader radio frequencies. If civilizations far more advanced than ours are, in principle, rarer, their technological resources should be more powerful. Hence the Soviet program, which looks at the Andromeda Galaxy two million light-years away.

What results have been obtained so far by observation? Here the greatest caution is imperative. Too many false alarms have been sounded, passed on, amplified and manipulated by news media always chasing the sensational in the area of science, and particularly in space. Remember the Soviet who started to observe the star that Kardashev had picked out as corresponding to his idea of interstellar communications. The laboratory's radio astronomer, Sholomitskii, observed this source, CTA 102 (California Technology Catalog A) and saw with his radio telescope that in the course of the year it oscillated regularly and sinusoidally with a periodicity of about seven days. At the time, this oscillation was incomprehensible; but since he was influenced by the work of Kardashev, he immediately assumed that there he had a serious indication of the artificial nature of this source; a source, he claimed, which was being modulated with a view to attracting the attention of possible observers.

Other radio astronomers, British and American in particular, immediately went to work to check Sholomitskii's observation. Unfortunately, they could not confirm the variations announced. At first, the Soviet replied that they had not operated in the same frequencies as he (810 megahertz instead of

900). Faced with incredulity, he eventually called a press conference in Moscow in 1964; *Pravda* immediately published a statement that a first sign of extraterrestrial intelligence had been detected.

Now, by a remarkable quirk of history, that same week at Mount Palomar in California a radio astronomer, of Dutch origin, Marteen Schmidt, studying CTA 102 from the standpoint of its optical spectrum, discovered that the spectral rays, incomprehensible at the time, were ordinary rays, of hydrogen for example, but shifted toward the red. Shifted to such a degree as to imply that the object was located far out in the depths of space. What was thought to be a harmless little star proved to be an ultrapowerful celestial body with a luminosity hundreds of times greater than that of a galaxy. He had discovered the first "quasar"! The Soviet's theory was thus routed at the very moment that it was brought with much fanfare to public knowledge. The *Astronomic Journal* published a letter from Schmidt describing his discovery only a few days after the *Pravda* article.

The lesson is worth remembering for the future. If we get alerts in the course of the SETI program, they must first be verified and corroborated by two or three telescopes before an announcement is made. The more so, since we must be suspicious of the inevitable hoaxes perpetrated by students at the top scientific schools, which even scientists are familiar with (and may relish) in this type of research. For this reason a small group was set up in collaboration with the International Institute of Space Law, the International Astronomical Federation, the International Academy of Astronautics, and the International Astronautical Federation. A procedure was worked out. In case of a discovery, it would be immediately verified by other radio telescopes. If the discovery were confirmed, a committee of experts of multidisciplinary composition, including biologists, would meet to decide whether or not to announce it. Once the decision were arrived at, the public announcement would be made by an international organization, undoubtedly the United Nations. This procedure would guarantee that the information was serious and would have symbolic value: if a signal is detected we want it to belong to all people, to become part of our intellectual heritage. No country, no institution, and no particular agency could appropriate it. These questions are now settled in principle; in two or three years everything will have been finalized.

There remains one more question, of quite another type. If we discover a signal, should we answer? What should we answer? And how? This has yet to be determined.

It will be the next chapter in the great adventure.

IX

THE ANTHROPIC
PRINCIPLE

Will we find a real terrestrial-type planet revolving around a star? Will we find amino acids in interstellar space, or adenine on Titan, or frozen bacteria on Mars? Will we pick up an artificial extraterrestrial signal? What melody will reach us then? Does the cosmic odyssey which has been, and remains, ours have any equivalents in other places or other times? Will we one day be enlightened by others for a deeper and more refined knowledge of the cosmos and a more precise knowledge of our own future?

The metaphysical scope of the research whose status and prospects I have just presented is as obvious as it is uplifting! Hence one should not be surprised by the heated philosophical discussions that have developed in recent years around it. One expression, intentionally coined, carries the entire intellectual and emotional weight of these discussions: the "anthropic" principle. Let me say straight away that in no sense did it inspire the SETI program. It is a speculative or philosophical principle which rests on an interpretation of the same facts on which our observational program is based, without really being linked to it. Many confusions and extrapolations arose from its announcement in 1974 by

Brandon Carter, the British astrophysicist at the Meudon Observatory. This announcement stated that: "The Universe must be such that it allows for the creation of observers within its midst at any stage." And he added, to emphasize the fundamental fact from which he started: "The existence of any organism which can be described as being an observer will be possible only for certain restricted combinations of parameters." Hence the formulation of the so-called "weak" version of the principle: "What we should expect to observe must be restricted by the conditions necessary to our presence as observers." In other words: Because we observe the cosmos, it must necessarily have been favorable to our appearance.

But from this weak version—which only registers the incontestable fact that for us human beings endowed with intelligence to be asking ourselves about the origins of the Universe a truly dizzying succession of favorable chances was needed—we have quickly gone on, sometimes without noticing, to a so-called "strong" version of the anthropic principle. From the term "chance," particularly when the chance is termed favorable, it is easy to slide into the notion of Providence. Thus, J. A. Wheeler, a powerful general relativity theorist, and others with him, have gone so far as to say that our existence is responsible for the special structure of the Universe. The emergence of the human species thus appears to be the goal, the destination or, if you will, the destiny of the Universe. If, despite its extremist finalism, the "strong" version of the anthropic principle has gained many adherents in recent years, it is because it fits rather easily into an interpretation, itself widespread, of the big bang theory. The 1980s have seen the emergence of new, very profound and very stimulating conceptions of what may have been the very first moments of the big bang, the inflationary big bang. These conceptions throw unexpected light on events which marked a primordial instant on which our destiny still depends fifteen billion years later; they decisively illuminate a dark area of the scenario which had been accepted by cosmology until then. It must be said that these views are still based on many speculative points, but they are sustained by an impressive theoretical framework borrowed from the two great physics theories of our century: the theory of general relativity and the theory of quantum mechanics. In fact, they appear as the magnificent fruit of an intense collaboration between astronomers concerned with cosmology and physicists concerned with particles. Their reasoning is highly technical and based on frequently very arduous mathematics; I will therefore confine myself to giving its general shape; we will see how it could have become linked with the "anthropic principle."

We know that physicists have demonstrated the existence of four fundamental interacting forces in nature: the strong interaction which, with prodigious intensity, links the component elements of atomic nuclei and ensures their cohesion; the weak interaction of very short range and minimal intensity manifested in the collision of certain particles (neutrinos, for example) and in certain nuclear reactions or disintegrations; the electromagnetic interaction representing the forces exerted through electric and magnetic fields between two bodies carrying electric charges; and gravitation. The question of the unification of these forces has been on the agenda for several decades. Spectacular success was initially achieved along this line of thought by quantum electrodynamics, when weak and electromagnetic interactions were unified. This theoretical success was crowned experimentally at CERN (the European Nuclear Research Center) in Geneva with the discovery of the W and Z bosons, particles whose existence had been predicted by theory. The physicists continued their work along the same path and succeeded in unifying these two interactions with the strong interaction. They thus built the "Unified Field Theory." This theory predicts the existence of new bosons, X bosons, with enormous masses, 10^{16} times that of protons, while the masses of the W and Z bosons are only one hundred times the mass of protons. It is out of the question to produce such particles in accelerators. Where are they to be found? It is here that we must turn to astrophysics. Just after the big bang, the Universe was so hot that such X bosons could have been present. The temperature was 10^{27} K and standard theory indicates that such a temperature could have become established 10^{-35} second after the big bang.

The consequences of the possible presence of these X bosons at that time should still be detectable today in the cosmos.

The current state of research allows us to present the following scenario for this ultrabrief but decisive period in the history of the cosmos: The three interactions became unified above 10^{27}K and separated below it. This implies, for reasons I cannot set forth in detail here, that perfect symmetry prevailed in this tiny, superdense cosmos before 10^{-35} second. And that this symmetry was then broken. What is called the "quantum vacuum" would have been symmetrical before that instant, and asymmetrical after it. Vacuum, they say, would then have undergone a "phase transition." It is pointless to say that vacuum in quantum mechanics must not be identified with nothingness. This vacuum has specific properties according to theory which, on this point, places it in an entirely different realm from common sense. One of the longest-known examples of these properties of vacuum is the sudden appearance, emanating from it, of a particle plus antiparticle pair with no energy input, by a mere random statistical fluctuation, with the only constraint of respecting Heisenberg's uncertainty principle.

This quantum "vacuum" is presented in the Unified Field Theory in such a way that in its symmetric state it has more energy than when the symmetry has been broken. It is also believed that the transitional phase may delay before it sets in. A convenient analogy may be given by using the example of another "phase transition" well known to everyone. When liquid water, which is symmetrical, freezes (below 0°C), a transition phase may intervene: the water is transformed into ice (with broken symmetry). But it turns out that, in the transition from the liquid state to the crystalline state, water passes from a high-energy state to a lower-energy state. The familiar proof of this difference in energy between the two phases of water is that in order to transform ice into liquid water, one must heat it. This is the principle of iceboxes: the ice which surrounds the food absorbs the heat coming from the outside, using it up to melt, and thus prevents the heat from reaching the food. Conversely, to transform liquid water into ice, energy must be taken from it. This time, we have the example of the icemaking machine. This difference in energy between the two states of the water comes from the fact that the molecules of water congealed in the regular arrangement of the crystal lattice interact with each other more strongly than when they are sliding over each other in the disorderly liquid state. This may be the case with the "quantum vacuum." Below the critical temperature of 10^{27}K, this vacuum could undergo a phase transition between a symmetrical state with a high energy level and a broken symmetry state at a lower energy level. Hence, when the vacuum goes from the first state to the second, it suddenly releases energy corresponding to the difference between the energies of the two states.

Something which is even more subtle: we know that, under certain conditions, we can cool water below 0°C without ice forming; what we then have is "supermelted" water, which changes suddenly into ice at the slightest shock. Hence we have a delayed "phase transition" where the water is liquid but charged with latent energy, ready to be released. Today the inclination is to believe that a phenomenon of this kind could have occured in the inflationary scenario of the cosmos. At 10^{-35} second, the vacuum must have had its symmetry broken for the reasons just stated. But that is not what happened. Each cubic yard of the expanding Universe would have been charged with the latent energy of the transitional phase. Let us call this energy "vacuum dynamite," for short.

We see that the standard big bang theory must be modified: it must give way to a far more rapid expansion process at such an exponentially high rate that it takes on a catastrophic course. This was true up to the end of the "supermelting" which, it is believed, occurred at 10^{-32} second. Within this unimaginably short interval between 10^{-35} and 10^{-32} second, space would have become 10^{50} times larger! Thus it is no exaggeration to speak of an "inflationary" period.

A new stage, a new consequence of these initial conditions: the sudden transition to broken symmetry instantaneously triggers the "vacuum dynamite." A colossal energy which immediately heats the Universe to 10^{27}K. Thus we witness a "second detonation" of the cosmos. A plethora of particles and antiparticles of all types is produced.

Since the Universe once more cools down under the effect of expansion (as currently accounted for by the standard theory), most of these particles and these antiparticles are annihilated by energy pairing and it is believed that this universal "cleansing" is achieved after one second, when the temperature drops back to its critical value of 10^{10}K. This is explained by statistical equilibrium reasoning. Now, nothing could have subsisted for the subsequent course of events and all matter and antimatter would have had to disappear if, by an extraordinary chance, there had not been at this precise moment a very slight excess of matter over antimatter. Because of this slight excess, when all the antimatter was annihilated with the portion of matter corresponding to it, a little matter was left over at the end of what can be called the "longest second" in history. Thus it is to the second bang that we owe the matter of which our world and ourselves are made; it is at this moment that, properly speaking, the grandiose fresco I have summarized above begins. These views remain theoretical, as I have stated, but they appear to be a promising hypothesis for research. Does this inflationary scenario reinforce, as some claim, the "strong" version of the anthropic principle? Does the slight excess of material that allowed the second bang to start the cosmic odyssey mark the intervention of some providence? I do not believe so; I think this scenario orients our thinking in quite a different direction.

Let us return to the end of the impressive inflationary "period": at this point the Universe was no larger than an apple, while to begin with it measured no more than 10^{-49} centimeters. Let us return to our analogy with the phase transition from liquid water to the crystalline state (ice). When one liter of "supercooled" water is transformed into ice, we know that not a single homogenous crystal emerges, but rather thousands of tiny crystals in a tangle, each homogeneous but randomly oriented. The same may have occurred with the vacuum: the vacuum cooled into a collection of homogeneous "crystals," each with "orientations" different from the broken symmetry.

Theoretical calculation indicates moreover that the "crystal" in which we are located has a size 10^{24} times larger than our observable Universe, of fifteen billion light-years.

We can see the first consequence: since our observable Universe is a very small part of a far larger homogeneous crystal, it must itself be homogeneous. This is indeed what has been observed for several decades and could never be explained by the big bang theory. The second consequence is that, if space was curved before the inflation, after its tremendous stretching, its radius of curvature should be immense; hence the curvature should be practically zero. This, too, has been observed and never explained.

Let us take this argument to its logical extreme. If this scenario is correct, space must be held to be divided into compartments: our Universe is a tiny crystal among other universes. Innumerable universes must be hypothesized. These universes could have properties different from ours. For example, the value e for the charge of an electron could have other values. Thus we may imagine different universes constituted in such a way that they could not have given rise to evolution leading to the appearance of beings like ourselves.

In this indefinite number of universes, there happens by chance to be one where the charge of the electron was such that excitation of the carbon atom was actually able to trigger the organic chemical processes I have described. And many other "miracles." But if the existing universes are so innumerable, why rule out the possibility that there may be others that benefitted from comparable chances? We can see that if the impressive cascade of chances, supported by the "weak" version of the "anthropic principle," gains in

110

credibility in the inflationary scenario, the "strong" anthropocentric version of this principle does not benefit from this progress.

If we throw our taboos, common sense, and prejudices overboard in order to come to know the Universe, be moved by its grandeur and enchanted with its beauty, humans appear not as the goal of the cosmic odyssey, beings who by their existence would unveil its meaning, but as the fruit, infinitely precarious and fragile, of a grandiose adventure whose destiny was highly whimsical, like a thin arabesque drawn on a frosted windowpane, a fragile line at the mercy of immense forces that overtake it and dispose of it, a light foam on the restless billows.

How can we not hope that tomorrow, by the force of their thoughts, these beings who have conquered the Earth, their planet, may—as they have always dreamed—pool their knowledge and their skills in the attempt to tame these billows and direct them for the good of all?

BIBLIOGRAPHY

BRACK, A., RAULIN, F. (ed.), "L'exobiologie" [Exo-biology], in *l'Astronomie*, Astronomical Society of France, December 1989.

FEINBERG, G., SHAPIRO, R., *Life Beyond Earth*, New York, Morrow and Co., 1980.

HEIDMANN, J., RIBES, J.-C., *A la recherche des extrater-restres* [Search for extraterrestrial life*], F. Nathan, 1985.

HEIDMANN, J., *L'Odyssée cosmique, quel destin pour l'univers?* [The cosmic odyssey, what fate for the Universe?*], Denoël, 1986.

HEIDMANN, J., *Dix années de recherches futurs des sig-naux extraterrestres* [Ten years of future search for extraterrestrial signals], in *Annales de Physique* 14, 133, French Society of Physics, 1989.

MARX, G. (ed.), *Bioastronomy: The Next Steps*, Kluwer Acad. Publ., 1988.

RAULIN, F., HEIDMANN, J., VINCENT, C., BRUNIER, S., *Dossier bioastronomie* [Bioastronomy dossier], in *Ciel et espace* [Sky and space*], French Association of Astronomy, June 1990.

TARTER, J. (ed.), "SETI—Search for Extraterrestrial Life," NP-114, Washington, NASA Headquarters, 1989.

---

* These references have not been published in English.